Lectures on

DIFFUSION PROBLEMS AND PARTIAL DIFFERENTIAL EQUATIONS

By

S. R. S. VARADHAN

Notes by

PL. MUTHURAMALINGAM
TARA R. NANDA

Published for the
TATA INSTITUTE OF FUNDAMENTAL RESEARCH, BOMBAY

SPRINGER-VERLAG
Berlin Heidelberg New York
1980

Author

S. R. S. VARADHAN
Courant Institute of Mathematical Sciences
New York University
251, Mercer Street
NEW YORK, N. Y. 10012.
U. S. A.

ISBN 3-540-08773-7. Springer-Verlag Berlin. Heidelberg. New York
ISBN 0-387-08773-7. Springer-Verlag New York. Heidelberg. Berlin

Printed by N. S. Ray at The BOOK CENTRE Limited
Sion East, Bombay 400 022 and published by H. Goetze
Springer-Verlag, Heidelberg, West Germany

PRINTED IN INDIA

CONTENTS

1.	The Heat Equation	...	1
2.	Kolmogorov's Theorem	...	11
3.	The One-dimensional Random Walk	...	14
4.	Construction of Wiener Measure	...	17
5.	Generalised Brownian Motion	...	31
6.	Markov Properties of Brownian Motion	...	33
7.	Reflection Principle	...	39
8.	Bluementhal's Zero-one Law	...	54
9.	Properties of Brownian Motion in One Dimension		57
10.	Dirichlet Problem and Brownian Motion	...	61
11.	Stochastic Integration	...	70
12.	Change of Variable Formula	...	89
13.	Extension to Vector-valued Ito Processes	...	92
14.	Brownian Motion as a Gaussian Process	...	98
15.	Equivalent form of Ito Process	...	102
16.	Ito's Formula	...	113
17.	Solution of Poisson's Equations	...	125
18.	The Feynman - Kac Formula	...	129
19.	An Application of the Feynman-Kac Formula (The Arc Sine Law)	...	134
20.	Brownian Motion with Drift	...	140
21.	Integral Equations	...	149
22.	Large Deviations	...	154

23. Stochastic Integral for a Wider Class of Functions ... 182

24. Explosions ... 189

25. Construction of a Diffusion Process ... 195

26. Uniqueness of Diffusion Process ... 205

27. On Lipschitz Square Roots ... 215

28. Random Time Changes ... 220

29. Cameron-Martin-Girsanov Formula ... 228

30. Behaviour of Diffusions for Large Times ... 235

31. Invariant Probability Distributions ... 242

32. Ergodic Theorem ... 266

33. Application of Stochastic Integral ... 272

APPENDIX

Language of Probability ... 275

Kolmogorov Theorem

Martingales

Uniform Integrability

Upcrossings and Down-crossings

BIBLIOGRAPHY ... 314

gk.

PREFACE

THESE ARE NOTES based on the lectures given at the T.I.F.R. Centre, Indian Institute of Science, Bangalore, during July and August of 1977. Starting from Brownian Motion, the lectures quickly got into the areas of Stochastic Differential Equations and Diffusion Theory. An attempt was made to introduce to the students diverse aspects of the theory. The last section on Martingales is based on some additional lectures given by K. Ramamurthy of the Indian Institute of Science. The author would like to express his appreciation of the efforts by Tara R. Nanda and PL. Muthuramalingam whose dedication and perseverance has made these notes possible.

S.R.S. Varadhan

1. THE HEAT EQUATION

LET US CONSIDER the equation

$$u_t - \frac{1}{2} \Delta u = 0 \tag{1}$$

which describes (in a suitable system of units) the temperature distribution of a certain honogeneous, isotropic body in the absence of any heat sources within the body. Here

$$u \equiv u(x_1, \ldots x_d, t); \; u_t \equiv \frac{\partial u}{\partial t}; \; \Delta u = \sum_{i=1}^{d} \frac{\partial^2 u}{\partial x_i^2},$$

t represents the time ranging over $[0,\infty)$ or $[0,T]$ and $x \equiv (x_1 \ldots x_d)$ belongs to $\mathbb{R}^d$.

We first consider the initial value problem. It consists in integrating equation (1) subject to the initial condition

$$u(o,x) = f(x). \tag{2}$$

The relation (2) is to be understood in the sense that

$$\underset{t \to 0}{\mathrm{Lt}} \; u(t,x) = f(x).$$

Physically (2) means that the distribution of temperature throughout the body is known at the initial moment of time.

We assume that the solution u has continuous derivatives, in the space coordinates upto second order inclusive and first order derivative in time.

It is easily verified that

$$u(t,x) = \frac{1}{(2\pi t)^{d/2}} \exp\left(-\frac{|x|^2}{2t}\right); \; |x|^2 = \sum_{i=1}^{d} x_i^2, \tag{3}$$

satisfies (1) and

(4) $$u(0,x) = \underset{t\to 0}{\mathrm{Lt}}\ u(t,x) = \delta(x)$$

Equation (4) gives us a very nice physical interpretation. The solution (3) can be interpreted as the temperature distribution within the body due to a unit source of heat specified at $t = 0$ at the space point $x = 0$. The linearity of the equation (1) now tells us that (by superposition) the solution of the initial value problem may be expected in the form

(5) $$u(t,x) = \int_{\mathbb{R}^d} f(y)\, p(t,x-y)dy,$$

where

$$p(t,x) = \frac{1}{(2\pi t)^{d/2}} \exp - \frac{|x|^2}{2t}.$$

<u>Exercise 1</u>. Let $f(x)$ be any bounded continuous function. Verify that $p(t,x)$ satisfies (1) and show that

(a) $\int p(t,x)\,dx = 1, \ \forall\ t > 0;$

(b) $\underset{t\to 0+}{\mathrm{Lt}} \int p(t,x)\, f(x)\, dx = f(0);$

(c) using (b) justify (4). Also show that (5) solves the initial value problem.

(Hints: For (a) use $\int_{-\infty}^{\infty} e^{-x^2} dx = \sqrt{\pi}$. For part (b) make the substitution $y = \frac{x}{\sqrt{\pi}}$ and apply Lebesgue dominated convergence theorem).

Since equation (1) is linear with constant coefficients it is invariant under time as well as space translations. This means that translates of solutions are also solutions. Further, for

$s \geq 0$, $t > 0$ and $y \in \mathbb{R}^d$,

$$(6) \qquad u(t,x) = \frac{1}{[2\pi(t+s)]^{d/2}} \exp - \frac{|x-y|^2}{2(t+s)}$$

and for $t > s$, $y \in \mathbb{R}^d$,

$$(7) \qquad u(t,x) = \frac{1}{[2\pi(t-s)]^{d/2}} \exp - \frac{|x-y|^2}{2(t-s)}$$

are also solutions of the heat equation (1).

The above method of solving the initial value problem is a sort of trial method, viz. we pick out a solution and verify that it satisfies (1). But one may ask, how does one obtain the solution? A partial clue to this is provided by the method of Fourier transforms. We pretend as if our solution $u(t,x)$ is going to be very well behaved and allow all operations performed on u to be legitimate.

Put $v(t,x) = \hat{u}(t,x)$ where ^ stands for the Fourier transform in the space variables only (in this case), i.e.

$$v(t,x) = \int_{\mathbb{R}^d} u(t,y)\, e^{i\, x-y}\, dy.$$

Using equation (1), one easily veri fies that

$$(8) \qquad v_t(t,x) = \frac{1}{2}|x|^2\, v(t,x)$$

with

$$(9) \qquad v(0,x) = \hat{f}(x).$$

The solution of equation (8) is given by

$$(10) \qquad v(t,x) = \hat{f}(x)\, e^{-t|x|^2/2}.$$

We have used (9) in obtaining (10).

Exercise 2. Verify that

$$\hat{p}(t,x) = \exp - \left(\frac{t|x|^2}{2}\right) .$$

Using Exercise 2, (10) can be written as

$$v(t,x) = \hat{u}(t,x) = \hat{f}(x)\hat{p}(t,x). \tag{11}$$

The right hand side above is the product of two Fourier transforms and we know that the Fourier transform of the convolution of two functions is given by the product of the Fourier transforms. Hence $u(t,x)$ is expected to be of the form (5).

Observe that if f is non-negative, then u is nonnegative and if f is bounded by M then u is also bounded by M in view of part (a) of Exercise 1.

THE INHOMOGENEOUS EQUATION. Consider the equation

$$v_t - \frac{\Delta v}{2} = g, \quad \text{with } v(0,x) = 0,$$

which describes the temperature within a homogeneous isotropic body in the presence of heat sources, specified as a function of time and space by $g(t,x)$. For $t > s$,

$$u(t,x) = \frac{1}{[2\pi(t-s)]^{d/2}} \exp - \frac{|x-y|^2}{2(t-s)}$$

is a solution of $u_t(t,x) - \frac{1}{2}\Delta u(t,x) = 0$ corresponding to a unit source at $t = s$, $x = y$. Consequently, a solution of the inhomogeneous problem is obtained by superposition.

Let

$$v(t,x) = \int_{\mathbb{R}^d} \int_0^t g(s,y)\, \frac{1}{[2\pi(t-s)]^{d/2}} \exp\left(- \frac{|x-y|^2}{2(t-s)}\right) dy\, ds$$

$$\text{i.e.}\quad v(t,x) = \int_0^t w(t,x,s)\,ds$$

where

$$w(t,x,s) = \int_{\mathbb{R}^d} g(s,y)\,\frac{1}{[2\pi(t-s)]^{d/2}}\exp\left(-\frac{|x-y|^2}{2(t-s)}\right)dy.$$

Exercise 3. Show that $v(t,x)$ defined above solves the inhomogeneous heat equation and satisfies $v(0,x) = 0$. Assume that g is sufficiently smooth and has compact support. $v_t - \frac{1}{2}\Delta v = \underset{s\to t}{\mathrm{Lt}}\, w(t,x,s)$ and now use part (b) of Exercise (1).

REMARK 1. We can assume g has compact support because in evaluating $v_t - \frac{1}{2}\Delta v$ the contribution to the integral is mainly from a small neighbourhood of the point (t,x). Outside this neighbourhood

$$\frac{1}{[2\pi(t-s)]^{d/2}}\exp\left(-\frac{|x-y|^2}{2(t-s)}\right)$$

satisfies

$$u_t - \frac{1}{2}\Delta u = 0.$$

2. If we put $g(s,y) = 0$ for $s < 0$, we recognize that $v(t,x) = g * p$. Taking spatial Fourier transforms this can be written as

$$v(t,\xi) = \int_0^t g(s,\xi)\exp -\tfrac{1}{2}(t-s)|\xi|^2 d\xi,$$

or

$$\frac{\partial \hat{v}}{\partial t} = \frac{\partial v}{\partial t} = g(t,\xi) + \frac{1}{2}\Delta v = (g(t,\xi) + \frac{1}{2}\Delta v)\ .$$

Therefore

$$\frac{\partial v}{\partial t} - \frac{1}{2}\Delta v = g.$$

Exercise 4. Solve $w_t - \frac{1}{2}\Delta w = g$ on $[0,\infty) \times \mathbb{R}^d$ with $w = f$ on $\{0\} \times \mathbb{R}^d$ (Cauchy problem for the heat equation).

Uniqueness. The solution of the Cauchy problem is unique provided the class of solutions is suitably restricted. The uniqueness of the solution is a consequence of the Maximum Principle.

MAXIMUM PRINCIPLE. *Let* u *be smooth and bounded on* $[0,T] \times \mathbb{R}^d$ *satisfying*

$$u_t - \frac{\Delta u}{2} \geq 0 \text{ in } (0,T] \times \mathbb{R}^d \text{ and } u(0,x) \geq 0, \forall x \in \mathbb{R}^d.$$

Then

$$u(t,x) \geq 0 \ \forall, t \in [0,T] \text{ and } \forall x \in \mathbb{R}^d.$$

Proof. The idea is to find minima for u or for an auxillary function.

Step 1. Let v be *any* function satisfying

$$v_t - \frac{\Delta v}{2} > 0 \text{ in } (0,T] \times \mathbb{R}^d.$$

Claim. v cannot attain a minimum for $t_o \in (0,T]$. Assume (to get a contradiction) that $v(t_o,x_o) \leq v(t,x)$ for some $t_o > 0$ and for all $t \in [0,T], \forall x \in \mathbb{R}^d$. At a minimum $v_t(t_o,x_o) \leq 0$, (since $t_o \neq 0$) $\Delta v(t_o,x_o) \geq 0$. Therefore

$$(v_t - \frac{\Delta v}{2})(t_o,x_o) \leq 0.$$

Thus, if v has any minimum it should occur at $t_o = 0$.

Step 2. Let $\varepsilon > 0$ be arbitrary. Choose α such that

$$h(t,x) = |x|^2 + \alpha t$$

satisfies

$$h_t - \frac{\Delta h}{2} = \alpha - d > 0 \quad (\text{say } \alpha = 2d).$$

Put $v_\varepsilon = u + \varepsilon h$. Then

$$\frac{\partial v_\varepsilon}{\partial t} - \frac{1}{2}\Delta v_\varepsilon > 0.$$

As u is bounded, $v_\varepsilon \to +\infty$ as $|x| \to +\infty$ v_ε must attain a minimum. This minimum occurs at $t = 0$ by Step 1. Therefore,

$$v_\varepsilon(t,x) \geq v_\varepsilon(0,x_o) \quad \text{for some } x_o \in \mathbb{R}^d,$$

i.e.

$$v_\varepsilon(t,x) \geq u(0,x_o) + \varepsilon|x_o|^2 > 0,$$

i.e.

$$u(t,x) + \varepsilon h(t,x) > 0, \forall \varepsilon.$$

This gives

$$u(t,x) \geq 0.$$

This completes the proof.

Exercise 5. (a) Let L be a linear differential operator satisfying $Lu = g$ on Ω (open in $\mathbb{R}^d$) and $u = f$ on $\partial\Omega$. Show that u is uniquely determined by f and g if and only if $Lu = 0$ on Ω and $u = 0$ on $\partial\Omega$ imply $u = 0$ on Ω.

(b) Let u be a bounded solution of the heat equation $u_t - \frac{1}{2}\Delta u = g$ with $u(0,x) = f(x)$. Use the maximum principle and part (a) to show that u is unique in the class of all bounded functions.

(c) Let

$$g(t) = \begin{cases} e^{-1/t^2}, & \text{if } t > 0, \\ 0, & \text{if } t \leq 0, \end{cases}$$

$$u(t,x) = \sum_{k=0}^{\infty} \frac{g^{(k)}(t/2)x^{2k}}{(2k)!}, \quad \text{on } R \times R.$$

Then

$$u(0,x) = 0, \quad u_t = \frac{\Delta u}{2}, \ u \not\equiv 0,$$

i.e. u satisfies

$$u_t - \frac{1}{2}\frac{\partial^2 u}{\partial x^2} = 0, \quad \text{with} \quad u(0,x) = 0.$$

This example shows that the solution is not unique because, u is not bounded. (This example is due to Tychonoff).

LEMMA 1. *Let* $p(t,x) = \frac{1}{(2\pi t)^{d/2}} \exp - \frac{|x|^2}{2t}$ *for* $t > 0$. *Then*

$$p(t,.)* p(s,.) = p(t+s,.).$$

Proof. Let f be any bounded continuous function and put

$$u(t,x) = \int_{\mathbb{R}^d} f(y)\, p(t,x-y)\, dy.$$

Then u satisfies

$$u_t - \frac{1}{2}\Delta u = 0, \ u(0,x) = f(x).$$

Let

$$v(t,x) = u(t+s,x).$$

Then

$$v_t - \frac{1}{2}\Delta v = 0, \ v(0,x) = u(s,x).$$

This has the unique solution

$$v(t,x) = \int u(s,y)\, p(t,x-y)\, dy.$$

Thus

$$\int_{\mathbb{R}^d} f(y)\, p(t+s,x-y)dy = \iint f(z)p(s,y-z)p(t,x-y)dzdy.$$

This is true for all f bounded and continuous. We conclude, therefore, that

$$p(t,.)* p(s,.) = p(t+s,.).$$

Exercise 6. Prove Lemma 1 directly using Fourier transforms.

It will be convenient to make a small change in notation which will be useful later on. We shall write $p(s,x,t,y) = p(t-s,y-x)$ for every x,y and $t > s$. $p(s,x,t,y)$ is called the *transition probability*, in dealing with Brownian motion. It represents the probability density that a "Brownian particle" located at space point x at time s moves to the space point y at a later time t.

NOTE. We use the same symbol p for the transition probability; it is a function of four variables and there will not be any ambiguity in using the same symbol p.

Exercise 7. Verify that

$$\int_{\mathbb{R}^d} p(s,x,t,y)p(t,y,\sigma,z)dy=p(s,x,\sigma,z), s < t < \sigma.$$

(Use Exercise 6).

REMARK. The significance of this result is obvious. The probability that the particle goes from x at time s to z at time σ is the sum total of the probabilities, that the particle moves from x at s to y at some intermediate time t and then to z at time σ.

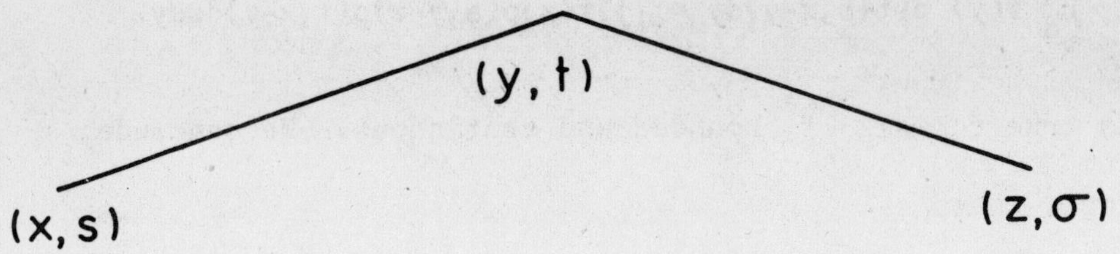

In this section we have introduced Brownian Motion corresponding to the operator $\frac{1}{2}\Delta$. Later on we shall introduce a more general diffusion process which corresponds to the operator $\frac{1}{2}\sum a_{ij} \frac{\partial^2}{\partial x_i \partial x_j} + \sum b_j \frac{\partial}{\partial x_j}$.

2. KOLMOGOROV'S THEOREM

DEFINITION. LET $(\Omega, \mathcal{B}, P)$ BE A probability space. A stochastic process in $\mathbb{R}^d$ is a collection $\{X_t : t \in I\}$ of $\mathbb{R}^d$-valued random variables defined on $(\Omega, \mathcal{B})$.

NOTE 1. I will always denote a subset of $\mathbb{R}^+ = [0,\infty)$.

2. X_t is also denoted by $X(t)$.

Let $\{X_t : t \in I\}$ be a stochastic process. For any collection $t_1, t_2, \ldots, t_k$ such that $t_i \in I$ and $0 \leq t_1 < t_2 < \ldots . < t_k$ and any Borel set Λ, in $\mathbb{R}^d \times \mathbb{R}^d \times \ldots \times \mathbb{R}^d$ (k times),, define

$$F_{t_1 \ldots t_k}(\Lambda) = P(w \in \Omega : (X_{t_1}(w), \ldots, X_{t_k}(w)) \in \Lambda).$$

If

$$\{t_1, \ldots, t_k\} \subset \{s_1, \ldots, s_\ell\} \subset I, \text{ with } l \geq k$$

such that

$$s_1^{(0)} < \ldots < s_{n_o}^{(0)} < t_1 < s_1^{(1)} \ldots < s_n^{(1)} < t_2 \ldots < t_k < s_1^{(k)} \ldots < s_{n_k}^{(k)},$$

let then

$$\Pi : \mathbb{R}^d \times \ldots \times \mathbb{R}^d (l \text{ times}) \to \mathbb{R}^d \times \ldots \times \mathbb{R}^d \ (k \text{ times})$$

be the canonical projection. If $E_{t_i} \subset \mathbb{R}^d$ is any Borel set in $\mathbb{R}^d$, $i = 1,2,\ldots k$, then

$$\Pi^{-1}(E_{t_1} \times \ldots \times E_{t_k}) = \mathbb{R}^d \times \ldots \times E_{t_1} \times \mathbb{R}^d \times \ldots \times E_{t_2} \times \ldots \times \mathbb{R}^d$$

(l times). The following condition always holds.

(*) $E_{t_1 \ldots t_k}(E_{t_1} \times \ldots \times E_{t_k}) = F_{s_1 \ldots s_l}(\Pi^{-1}(E_{t_1} \times \ldots \times E_{t_k}))$.

If (*) holds for an arbitrary collection $\{F_{t_1} \ldots t_k : 0 \leq t_1 < t_2 \ldots < t_k\}$ $(k = 1,2,3\ldots)$ of ditributions then it is said to satisfy the *consistency condition.*

Exercise 1. (a) Verify that $F_{t_1} \ldots t_k$ is a probability measure on $\mathbb{R}^d \times \ldots \times \mathbb{R}^d$ (k times).

(b) Verify (*). (If B_m denotes the Borel σ field of $\mathbb{R}^m$, $B_{m+n} = B_m \times B_n$).

The following theorem is a converse of Exercise 1 and is often used to identify a stochastic process with a family of distributions satisfying the consistency condition.

KOLMOGOROV's THEOREM.

Let $\{F_{t_1,t_2,\ldots t_k}$ $0 \leq t_1 < t_2 < \ldots < t_k < \infty\}$ *be a family of probability distributions (on* $\mathbb{R}^d \times \ldots \times \mathbb{R}^d$, *k times,* $k = 1,2\ldots$*) satisfying the consistency condition. Then there exists a measurable space* $(\Omega_K, \mathcal{B})$, *a unique probability measure* P *on* $(\Omega_K, \mathcal{B})$ *and a stochastic process* $\{X_t : 0 \leq t < \infty\}$ *such that the family of probability measures associated with it is precisely*

$$\{F_{t_1,t_2,\ldots t_k} : 0 \leq t_1 < t_2 < \ldots < t_k < \infty\}, \quad k = 1,2,\ldots .$$

A proof can be found in the APPENDIX. We mention a few points about the proof which prove to be very useful and should be observed carefully.

1. The space Ω_K is the set of all $\mathbb{R}^d$ -valued functions defined on $[0,\infty)$:

$$\Omega_K = \prod_{t \in [0,\infty)} \mathbb{R}^d$$

2. The random variable X_t is the tth -projection of Ω_K onto $\mathbb{R}^d_t$.

3. $\mathcal{B}$ is the smallest σ -algebra with respect to which all the projections are measurable.

4. P given by

$$P(w:X_{t_1}(w) \in A_1,\ldots X_{t_k}(w) \in A_k) = F_{t_1\ldots t_k}(A_1 \times\ldots\times A_k)$$

where A_i is a Borel set in $\mathbb{R}^d$, is a measure on the algebra generated by $\{X_{t_1},\ldots X_{t_k}\}$ $(k = 1,2,3\ldots)$ and extends uniquely to $\mathcal{B}$.

<u>REMARK</u>. Although the proof of Kolmogorov's theorem is very constructive the space Ω_K is too "large" and the σ -algebra $\mathcal{B}$ too "small" for practical purposes. In applications one needs a "nice" collection of $\mathbb{R}^d$ -valued functions (for example continuous, or differentiable functions), a "large" σ -algebra on this collection and a probability measure concentrated on this family .

3. THE ONE DIMENSIONAL RANDOM WALK

BEFORE WE TAKE up Brownian motion, we describe a one dimensional random walk which in a certain limiting case possesses the properties of Brownian motion.

Imagine a person at the position $x = 0$ at time $t = 0$. Assume that at equal intervals of time $t = \tau$ he takes a step either along the positive x axis or the negative x axis and reaches the point $x(t) = x(t-\tau)+ h$ or $x(t) = x(t-\tau)- h$ respectively. The probability that he takes a step in either direction is assumed to be 1/2. Denote by $f(x,t)$ the probability that after the time $t = n\tau$ (n intervals of time τ) he reaches the position x. If he takes m steps to the right (positive x-axis) in reaching x then there are ${}^nC_m$ possible ways in which he can achieve these m steps. Therefore, the probability $f(x,t)$ is ${}^nC_m(\frac{1}{2})^n$.

$f(x,t)$ satisfies the difference equation

$$f(x,t+\tau) = \frac{1}{2} f(x-h,t) + \frac{1}{2} f(x+h,t) \tag{1}$$

and

$$x = h(m-(n-m)) = (2m-n)h. \tag{2}$$

To see this one need only observe that to reach $(x,t+\tau)$ there are two ways possible, viz. $(x-h,t) \to (x,t+\tau)$ or $(x+h,t) \to (x,t+\tau)$ and the probability for each one of these is 1/2. Also note that by definition of f,

(3) $$f(h,\tau) = \frac{1}{2} = f(-h,\tau),$$

so that

(4) $$f(x,t+\tau) = f(h,\tau)f(x-h,t) + f(-h,\tau)f(x+h,t).$$

The reader can identify (4) as a "discrete version" of convolution.

By our assumption,

(5) $$f(0,0) = 1, \quad f(x,0) = 0 \quad \text{if} \quad x \neq 0.$$

We examine equation (1) in the limit $h \to 0$, $\tau \to 0$. To obtain reasonable results we cannot let h and τ trend to zero arbitratily. Instead we assume that

(6) $$\frac{h}{\tau} \to 1 \quad \text{as} \quad h \to 0 \quad \text{and} \quad \tau \to 0.$$

The physical nature of the problem suggests that (6) should hold. To see this we argue as follows. Since the person is equally likely to go in either direction the average value of x will be 0. Therefore a reasonable measure of the "progress" made by the person is either $|x|$ or x^2. Indeed, since x is a random variable (since m is one) one gets, using (2),

$$E(x) = 2E(m) - n = 0, \quad E(x^2) = h^2E((2m-n)^2) = h^2n.$$

(Use $\sum_{m=0}^{n} m \; {}^{n}C_m (\frac{1}{2})^n = \frac{n}{2}$, $\sum_{m=0}^{n} m^2 \; {}^{n}C_m (\frac{1}{2})^n = \frac{n(n+1)}{4}$).

Thus

$$E\{\frac{x^2}{t}\} = \frac{1}{t} E(x^2) = \frac{h^2n}{n\tau} = \frac{h^2}{\tau},$$

and as t becomes large we expect that the average distance covered per unit time remains constant. (This constant is chosen to be 1 for reasons that will become apparent later). This justifies (6). In fact, a simple argument shows that if $\frac{h^2}{\tau} \to 0$ or $+\infty$, x may approach $+\infty$ in a finite time which is physically untenable.

(1) now gives

$$f(x,t+\tau) - f(x,t) = \frac{1}{2}\{f(x-h,t)-f(x,t)+f(x+h,t)-f(x,t)\} .$$

Assuming sufficient smoothness on f, we get in the limit as $h,\tau \to 0$ and in view of (6),

$$\frac{\partial f}{\partial t} = \frac{1}{2}\frac{\partial^2 f}{\partial x^2} \tag{7}$$

(to get the factor 1/2 we choose $\frac{h^2}{\tau} \to 1$). This is the equation satisfied by the probability density f. The particle in this limit performs what is known as *Brownian motion* to which we now turn our attention.

References. 1. GNEDENKO: *The theory of probability*, Ch.10.

2. *The Feynman Lectures on physics*, Vol.1, Ch.6.

4. CONSTRUCTION OF WIENER MEASURE

ONE EXAMPLE WHERE the Kolmogorov construction yields a probability measure concentrated on a "nice" class Ω is the Brownian motion.

DEFINITION. A Brownian motion with starting point x is an $\mathbb{R}^d$-valued stochastic process $\{X(t): 0 \leq t < \infty\}$ where

(i) $X(0) = x$ = constant;

(ii) the family of distribution is specified by

$$F_{t_1}\dots_{t_k}(A) = \int_A p(0,x,t_1,x_1)p(t_1,x_1,t_2,x_2)\dots p(t_{k-1},x_{k-1},t_k,x_k)dx_1\dots dx_k$$

for every Borel set A in $\mathbb{R}^d \times\dots\times \mathbb{R}^d$ (k times).

N.B. The stochastic process appearing in the definition above is the one given by the Kolmogorov construction.

It may be useful to have the following picture of a Brownian motion. The space Ω_K may be thought of as representing particles performing Brownian movement; $\{X_t: 0 \leq t < \infty\}$ then represents the trajectories of these particles in the space $\mathbb{R}^d$ as functions of time and $\mathcal{B}$ can be considered as a representation of the observations made on these particles.

Exercise 2. (a). Show that $F_{t_1\dots t_k}$ defined above is a probability measure on $\mathbb{R}^d \times\dots\times \mathbb{R}^d$ (k times).

(b) $\{F_{t_1\dots t_k} : 0 \leq t_1 < t_2 <\dots< t_k < \infty\}$ satisfies the consistency condition. (Use Fubini's theorem).

(c) $X_{t_1} - x, X_t - X_{t_1}, \ldots, X_{t_k} - X_{t_{k-1}}$ are independent random variables and if $t > s$, then $X_t - X_s$ is a random variable whose distribution density is given by

$$p(t-s,y) = \frac{1}{[2\pi(t-s)]^{d/2}} \exp\left(-\frac{1}{2}(t-s)^{-1}|y|^2\right).$$

(Hint: Try to show that $X_{t_1} - x, X_{t_2} - X_{t_1}, \ldots, X_{t_k} - X_{t_{k-1}}$ have a joint distribution given by a product measure. For this let ϕ be any bounded real measurable function on $\mathbb{R}^d \times \ldots \times \mathbb{R}^d$ (k times). Then

$$E_{X_{t_1}-x,\, X_{t_2}-X_{t_1}, \ldots, X_{t_k}-X_{t_{k-1}}}(\phi(Z_1,\ldots,Z_k)) = E_{X_{t_1},\ldots,X_{t_k}}(\phi(Z_1 - x, \ldots, Z_k - Z_{k-1}))$$

where $E_{X_{t_1} \ldots X_{t_k}}(\phi)$ is the expectation of ϕ with respect to the joint distribution of $(X_{t_1} \ldots, X_{t_k})$. You may also require the change of variable formula).

<u>Problem</u>. Given a Brownian motion with starting point x our aim is to find a probability P_x on the space $\Omega = C([0,\infty); \mathbb{R}^d)$ of all continuous functions from $[0,\infty) \to \mathbb{R}^d$ which induces the Brownian motion. We will thus have a ***continuous realisation*** to Brownian motion. To achieve this goal we will work with the collection $\{F_{t_1,\ldots t_k} : 0 \leq t_1 < t_2 < \ldots < t_k\}$ where $t_i \in D$, a countable dense subset of $[0,\infty)$.

Step 1. The first step is to find a probability measure on a "smaller" space and lift it to $C([0,\infty);\mathbb{R}^d)$. Let

$$\Omega = C([0,\infty);\mathbb{R}^d),$$

D a countable dense subset of $[0,\infty)$; $\Omega(D) = \{f: D \to \mathbb{R}^d\}$ where f is uniformly continuous on $[0,N] \cap D$ for $N = 1,2,\ldots$. We equip Ω with the topology of uniform convergence on compact sets and $\Omega(D)$ with the topology of uniform convergence on sets of the form $D \cap K$ where $K \subset [0,\infty)$ is compact; Ω and $\Omega(D)$ are separable metric spaces isometric to each other.

Exercise 3. Let

$$p_n(f,g) = \sup_{0 \le t \le n} |f(t)-g(t)| \quad \text{for } f,g \in \Omega$$

and

$$p_{n,D}(f,g) = \sup_{\substack{0 \le t \le n \\ t \in D}} |f(t)-g(t)| \quad \text{for } f,g \in \Omega(D).$$

Define

$$\rho(f,g) = \sum_{n=1}^{\infty} \frac{1}{2^n} \frac{p_n(f,g)}{1+p_n(f,g)}, \ \forall\, f,g \in \Omega,$$

$$\rho_D(f,g) = \sum_{n=1}^{\infty} \frac{1}{2^n} \frac{p_{n,D}(f,g)}{1+p_{n,D}(f,g)}, \ \forall\, f,g \in \Omega(D).$$

Show that

(i) $\{f_n\} \subset \Omega$ converges to f if and only if $f_n \to f$ uniformly on compact subsets of $[0,\infty)$;

(ii) $\{f_n\} \subset \Omega(D)$ converges to f if and only if $f_n|_{D\cap K} \to f|_{D\cap K}$ uniformly for every compact subset K of $[0,\infty)$;

(iii) $\{(P_1,\ldots,P_d)\}$ where P_i is a polynomial with rational coefficients is a countable dense subset of Ω;

(iv) $\{(P_{1\,D},\ldots,P_{d\,D})\}$ is a countable dense subset of $\Omega(D)$;

(v) $\tau: \Omega \to \Omega(D)$ where $\tau(f) = f|_D$ is a (ρ,ρ_D)-isometry of Ω onto $\Omega(D)$;

(vi) if $V(f,\varepsilon,n) = \{g \in \Omega : p_n(f,g) < \varepsilon\}$ for $f \in \Omega$, $\varepsilon > 0$ and

$$V_D(f,\varepsilon,n) = \{g \in \Omega(D) : p_{n,D}(f,g) < \varepsilon\} \text{ for } f \in \Omega(D), \varepsilon > 0,$$

then

$$\{V(f,\varepsilon,n) : f \in \Omega, \varepsilon > 0, n = 1,2\ldots\}$$

is a base for the topology of Ω and

$$\{V_D(f,\varepsilon,n) : f \in \Omega(D), \varepsilon > 0, n = 1,2,\ldots\}$$

is a base for the topology of $\Omega(D)$.

REMARK. By Exercise 3(v) any Borel probability measure on $\Omega(D)$ can be lifted to a Borel probability measure on Ω.

2nd Step. Define the modulus of continuity $\Delta_D^{T,\delta}(f)$ of a function f on D in the interval $[0,T]$ by

$$\Delta_D^{T,\delta}(f) = \sup\{|f(t)-f(s)| : |t-s| < \delta\ t,s \in D \cap [0,T]\}$$

As D is countable one has

Exercise 4. (a) Show that $f:\Delta_D^{N,\frac{1}{j}}(f) \leq \frac{1}{k}\}$ is measurable in the σ-algebra generated by the projections

$$\Pi_t : \Pi\{\mathbb{R}_t^d : t \in D\} \to \mathbb{R}_t^d$$

Proof. The lemma is equivalent to showing that $\mathcal{B} = \sigma(\mathcal{E})$. As each of the projection $\Pi_{t_1 \ldots t_k}$ is continuous, $\sigma(\mathcal{E}) \subset \mathcal{B}$. To show that $\mathcal{B} \subset \sigma(\mathcal{E})$, it is enough to show that $V_D(f,\varepsilon,n) \in \mathcal{E}$ because $\Omega(D)$ is separable. (Cf. Exercise 3(iv) and 3(vi)). By definition

$$V_D\,(f,\epsilon,n) = \{g \in \Omega(D) : p_{n,D}(f,g) < \epsilon\}$$
$$= \bigcup_{m=1}^{\infty} \{g \in \Omega(D) : p_{n,D}(f,g) \leq \varepsilon - \frac{1}{m}\}$$
$$= \bigcup_{m=1}^{\infty} \{g \in \Omega(D) : |g(t_i) - f(t_i)| \leq \epsilon - \frac{1}{m}, \forall\ t_i \in D \cap [0,n]\} .$$

The result follows if one observes that each Π_{t_i} is continuous.

REMARK 1. The lemma together with Exercise 4(b) signifies that the Kolmogorov probability P_x on $\Pi\{\mathbb{R}^d_t : t \in D\}$ is defined on the topological Borel σ-field of $\Omega(D)$.

2. The proof of the lemma goes through if $\Omega(D)$ is replaced by Ω.

Step 3. We want to show that $P_x(\Omega(D)) = 1$. By Exercise 4(b) this is equivalent to showing that $\operatorname*{Lt}_{j \to \infty} P(\Delta_D^{N,1/j}(f) \leq \frac{1}{k}) = 1$ for all N and k. The lemmas which follow will give the desired result.

LEMMA (Lévy). *Let* $X_1, \ldots X_n$ *be independent random variables,* $\varepsilon > 0$ *and* $\delta > 0$ *arbitrary. If*

$$P(|X_r + X_{r+1} + \ldots + X_\ell| \geq \delta) \leq \varepsilon$$

$\forall\ r,\ell$ *such that* $1 \leq r \leq \ell \leq n$, *then*

$$P(\sup_{1 \leq j \leq n} |X_1 + \ldots + X_j| \geq 2\delta) \leq 2\varepsilon.$$

(see Kolmogorov's theorem) for every $j = 1,2,\ldots,$ for every $N = 1,2,\ldots$ and for every $k = 1,2\ldots$. (Hint: Use the fact that the projections are continuous).

(b) Show that $\Omega(D) = \bigcap_{N=1}^{\infty} \bigcap_{k=1}^{\infty} \bigcup_{j=1}^{\infty} \{\Delta_D^{N,\frac{1}{j}}(f) \leq \frac{1}{k}\}$

and hence $\Omega(D)$ is measurable in $\Pi\{\mathbb{R}_t^d : t \in D\}$.

Let $\Pi_{t_1\ldots t_k} : \Omega(D) \to \mathbb{R}^d \times \mathbb{R}^d \times\ldots\times \mathbb{R}^d$ (k times) be the projections and let

$$\mathcal{E}_{t_1\ldots t_k} = \Pi^{-1}_{t_1\ldots t_k}(\underbrace{(\mathbb{R}^d)\times\ldots\times(\mathbb{R}^d)}_{k \text{ times}}).$$

Put

$$\mathcal{E} = \bigcup\{\mathcal{E}_{t_1\ldots t_k} : 0 \leq t_1 < t_2 <\ldots< t_k < \infty;\ t_i \in D\} .$$

Then, as

$$\mathcal{E}_{t_1\ldots t_k} \cup \mathcal{E}_{s_1\ldots s_l} \subset \mathcal{E}_{\tau_1\ldots\tau_m},$$

where

$$\{t_1\ldots t_k, s_1\ldots s_l\} \subset \{\tau_1\ldots\tau_m\} ,$$

$\mathcal{E}$ is an algebra. Let $\sigma(\mathcal{E})$ be the σ -algebra generated by $\mathcal{E}$.

<u>LEMMA</u>. *Let $\mathcal{B}$ be the (topological) Borel σ -field of $\Omega(D)$. Then $\mathcal{B}$ is the σ -algebra generated by all the projections*

$\{\Pi_{t_i\ldots t_k} : 0 \leq t_1 < t_2 < \quad < t_k, t_i \in D\}$.

REMARK. By subadditivity it is clear that

$$P\left(\sup_{1 \leq j \leq n} |X_1 + \ldots + X_j| \geq 2\delta\right) \leq n\varepsilon .$$

Ultimately, we shall let $n \to \infty$ and this estimate is of no utility. The importance of the lemma is that it gives an estimate independent of n.

Proof. Let $S_j = X_1 + \ldots + X_j$, $E = \left\{ \sup_{1 \leq j \leq n} |S_j| \geq 2\delta \right\}$. Put

$$E_1 = \{|S_1| \geq 2\delta\},$$

$$E_2 = \{|S_1| < 2\delta,\ |S_2| \geq 2\delta\},$$

$$\ldots \quad \ldots \quad \ldots \quad \ldots$$

$$\ldots \quad \ldots \quad \ldots \quad \ldots$$

$$E_n = \{|S_j| < 2\delta,\ 1 \leq j \leq n-1,\ |S_n| \geq 2\delta\}.$$

Then

$$E = \bigcup_{j=1}^{n} E_j, \quad E_j \cap E_i = \phi \ \text{ if } \ j \neq i;$$

$$P\{E \cap (|S_n| \leq \delta) = P\left(\bigcup_{j=1}^{n} (E_j \cap (|S_n| \leq \delta)\right)$$

$$\leq P\left\{\bigcup (E_i \cap (|S_n - S_j| \geq \delta))\right\}$$

$$\leq \sum_{j=1}^{n} P(E_j)\, P(|S_n - S_j| \geq \delta) \text{ (by independence)}$$

$$\leq \varepsilon P(E) \quad \text{(b y data).}$$

$$P\{E \cap (|S_n| > \delta)\} \leq P(|S_n| > \delta) \leq \varepsilon \quad \text{(by data).}$$

Combining the two estimates above, we get

$$P(E) \leq \varepsilon + \varepsilon P(E).$$

If $\varepsilon > \frac{1}{2}$, $2\varepsilon > 1$. If $\varepsilon < \frac{1}{2}$, $\frac{\varepsilon}{1-\varepsilon} \leq 2\varepsilon$. I n either case $P(E) \leq 2\varepsilon$.

LEMMA. *Let* $\{X(t)\}$ *be a Brownian motion,* $I \subset [0,\infty)$ *be a finite interval,* $F \subset I \cap D$ *be finite. Then*

$$P_x(\sup_{t,\sigma \in F} |X(t)-X(\sigma)| \geq 4\delta) \leq C(d) \frac{|I|^2}{\delta^4},$$

where $|I|$ *is the length of the interval and* $C(d)$ *a constant depending only on* d.

REMARK. Observe that the estimate is independent of the finite set F.

F

Proof. Let $F = \{t_i: 0 \leq t_1 < t_2 < \ldots < t_k < \infty\}$.

Put

$$X_1 = X(t_2) - X(t_1), \ldots, X_{k-1} = X(t_k) - X(t_{k-1}).$$

Then $X_1 \ldots X_{k-1}$ are independent (Cf. Exercise 2(c)). Let

$$\varepsilon = \sup_{1 \leq r \leq l \leq k-1} P_x(|X_r + X_{r+1} + \ldots + X_l| \geq \delta).$$

Note that

$$P_x(|X_r + \ldots + X_l| \geq \delta) = P(|X(t')-X(t'')| \geq \delta) \text{ for some } t', t'' \text{ in } F$$

$$\leq \frac{E(|X(t')-X(t'')|^4)}{\delta^4} \quad \text{(see Tchebyshey's inequality in Appendix)}$$

$$(*) \qquad \leq \frac{C'(t''-t')}{\delta^4} \quad (C' = \text{constant})$$

$$\leq \frac{C'|I|^2}{\delta^4}.$$

Therefore $\varepsilon \leq \frac{C'|I|^2}{\delta^4}$. Now

$$P_x\Big(\sup_{t,\sigma \in F} |X(t) - X(\sigma)| \geq 4\delta\Big)$$

$$P_x\Big(\sup_{1 \leq i \leq k} |X(t_i)-X(t_1)| \geq 2\delta\Big)$$

$$= P_x\Big(\sup_{i \leq j \leq k-1} |X_1+\ldots+ X_j| \geq 2 \Big)\leq 2\varepsilon \quad \text{(by previous lemma)}$$

$$\frac{2C'|I|^2}{\delta^4} = \frac{C|I|^2}{\delta^4} .$$

Exercise 5. Verify (*).

(Hint: Use the density function obtained in Exercise 2(c) to evaluate the expectation and go over to "popular" coordinates. (The value of C' is $d(2d+1)$)).

LEMMA. $$P_x\Big\{ \sup_{\substack{|t-s| \leq h \\ t,s \in [0,T] \cap D}} |X(t)-X(s)| > \rho \Big\} = P_x(\Delta_D^{T,h} > \rho)$$

$$\leq \Phi (T,\rho,h) = C \frac{h}{\rho^4} \Big(\Big[\frac{T}{h}\Big] + 1\Big).$$

Note that $\Phi(T,\rho,h) \to 0$ as $h \to 0$ for every fixed T and ρ.

Proof. Define the intervals $I_1,I_2,\ldots$ by

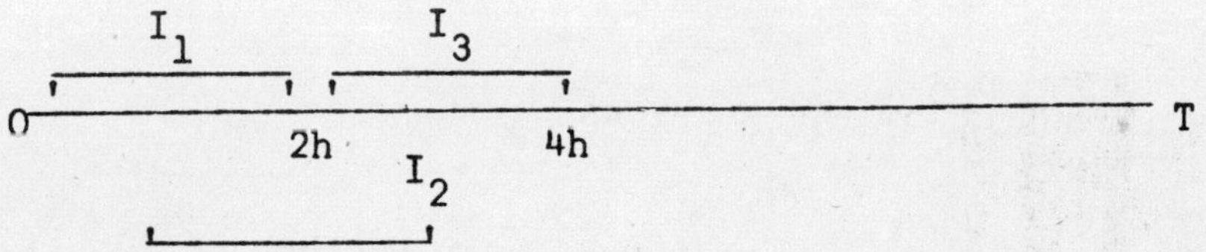

$$I_k = [(k-1)h,(k+1)h] \cap (0,T], \quad k = 1,2,\ldots .$$

Let $I_1,I_2,\ldots I_r$ be those intervals for which

$$I_j \cap [0,T] \neq \phi \ (j = 1,2,\ldots,r).$$

Clearly there are $[\frac{T}{h}] + 1$ of them. If $|t-s| \le h$ then $t,s \in I_j$ for some j, $1 \le j \le r$. Write $D = \bigcup_{n=1}^{\infty} F_n$ where $F_n \subset F_{n+1}$ and F_n is finite. Then

$$P_x\{ \sup_{\substack{|t-s| \le h \\ t,s \in [0,T] \cap D}} |X(t)-X(s)| > \rho\} = P_x\{ \bigcup_{n=1}^{\infty} (\sup_{\substack{|t-s| \le h \\ t,s \in D \cap F_n}} |X(t)-X(s)| > \rho)\}$$

$$= \sup_n P_x \{\sup_j \sup_{t,s \in F_n \cap I_j} (|X(t)-X(s)| > \rho)\}$$

$$\le \sup_n \sum_{j=1}^{r} P_x (\sup_{t,s \in F_n \cap I_j} (|X(t)-X(s)| > \rho))$$

$$\le \sup_n ([\frac{T}{h}] + 1) \frac{C(2h)^2}{(\rho/4)^4} \text{ by the last lemma}$$

$$\le \Phi (T,\rho,h).$$

THEOREM. $P_x(\Omega(D)) = 1$.

Proof. It is enough to show that

$$\operatorname*{Lt}_{j \to \infty} P_x (\Delta_D^{N,\frac{1}{j}} (f) \le \frac{1}{k}) = 1 \quad \text{(See Exercise 4(b))}.$$

But this is guaranteed by the previous lemma.

REMARK. 1. It can be shown that the outer measure of Ω is 1.

2. Ω is not measurable in $\prod_{t \ge 0} \mathbb{R}_t^d$.

Let $\tilde{P}_x$ be the measure on Ω induced by P_x on $\Omega(D)$. We have already remarked that P_x is defined on the (topological Borel σ field of $\Omega(D)$. As P_x is a probability measure, $\tilde{P}_x$ is also a probability measure. It should now be verified that $\tilde{P}_x$ is really the probability measure consistent with the given distribution.

THEOREM. $\tilde{P}_x$ *is a required probability measure for a continuous realization of the Brownian motion.*

Proof. We have to show that

$$F_{t_1 \ldots t_k} = \tilde{P}_x \Pi^{-1}_{t_1 \ldots t_k} \quad \text{for all } t_1, t_2 \ldots t_k \text{ in } [0,\infty).$$

Step 1. Let $t_1, \ldots, t_k \in D$. Then

$$P_x(\Pi^{-1}_{t_1 \ldots t_k}(A_1 \times, \ldots \times A_k)) = P_x(\tau\Pi^{-1}_{t_1 \ldots t_k}(A_1 \times \ldots \times A_k))$$

for every A_i Borel in $\mathbb{R}^d$. The right side above is

$$P_x(\Pi^{-1}_{t_1 \ldots t_k}(A_1 \times \ldots \times A_k)) = F_{t_1 \ldots t_k}(A_1 \times \ldots \times A_k)$$

(by definition of P_x).

Step 2. We know that $T_{t_1, t_2 \ldots t_k} = \tilde{P}_x \Pi_{t_1, t_2, \ldots, t_k}$ provided that $t_1, t_2, \ldots, t_k \in D$. Let us pick $t_1^{(n)}, \ldots, t_k^{(n)}$, such that each $t_i^{(n)} \in D$ and $t_k^{(n)} \to t_k$ as $n \to \infty$. For each n and for each fixed $f\colon \mathbb{R}^d \to \mathbb{R}$ which is bounded and continuous,

$$E^{F_{t_1^{(n)}, \ldots t_k^{(n)}}}[f(x_1, \ldots, x_k)] = E^{\tilde{P}_x}[f(x(t_1^{(n)}, \ldots, x(t_k^{(n)})].$$

Letting $n \to \infty$ we see that

$$E^{F_{t_1,\ldots,t_k}}[f(x_1,\ldots,x_k)] = E^{P_x}[f(x(t_1),\ldots,x(t_k))]$$

for all $t_1,\ldots t_k$. This completes the proof.

The definition of the Brownian motion given earlier has a built-in constraint that all "trajectories" start from $X(0) = x$. This result is given by

THEOREM. $\tilde{P}_o\{f: f(0) = 0\} = 1.$

Proof. Obvious; because $E^{\tilde{P}_x}[\phi(x(0))] = \phi(x)$.

NOTE. In future $\tilde{P}_x$ will be replaced by P_x and $\tilde{P}_o = P_o$ will be denoted by P.

Let $T_x: \Omega \to \Omega$ be the map given by $(T_x f)(t) = x + f(t)$. T_x translates every 'trajectory' through the vector x.

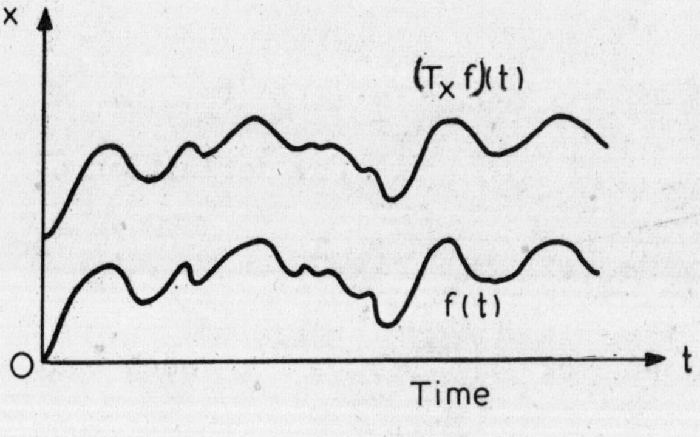

Let us conceive a real Brownian motion of a system of particles. The operation T_x means that the system is translated in space (along with everything else that affects it) through a vector x. The symmetry of the physical laws governing this motion tells us that any

property exhibited by the first process sh ould be exhibited by the second process and vice versa. Mathematically this is expressed by

<u>THEOREM</u>. $$P_x = PT_x^{-1} .$$

<u>Proof</u>. It is enough to show that

$$P_x(T_x \Pi^{-1}_{t_1\ldots t_k}(A_1 \times\ldots\times A_k))=P(\Pi^{-1}_{t_1\ldots t_k}(A_1 \times\ldots\times A_k))$$

for every A_i Borel in $\mathbb{R}^d$. Clearly,

$$T_x \Pi^{-1}_{t_1\ldots t_k}(A_1 \times\ldots\times A_k)=\Pi^{-1}_{t_1\ldots t_k}(A_1-x \times\ldots\times A_k-x).$$

Thus we have only to show that

$$\int_{A_1-x}\int\ldots\int_{A_k-x} p(0,x,t_1,x_1)\ldots p(t_{k-1},x_{k-1},t_k,x_k)dx_1\ldots dx_k$$

$$= \int_{A_1}\ldots\int_{A_k} p(0,0,t_1,x_1)\ldots p(t_{k-1},x_{k-1},t_k,x_k)dx_1\ldots dx_k,$$

which is obvious.

<u>Exercise</u>. (a) If $\beta(t,.)$ is a Brownian motion s tarting at $(0,0)$ then $\frac{1}{\sqrt{\varepsilon}}\beta(\varepsilon t)$ is a Brownian motion starting at $(0,0)$ for every $\varepsilon > 0$.

(b) If X is a d-dimensional Brownian motion and Y iss a d'-dimensional Brownian motion then (X,Y) is a $d+d'$ dimensional Brownian motion provided that X and Y are independent.

(c) If $X_t = (X_t^1,\ldots,X_t^d)$ is a d-dimensional Brownian motion, then X_t^j is a one-dimensional Brownian motion. $(j = 1,2,\ldots d)$.

$$\tau(w) = \inf\{t: |X_t(w)| \geq +1\}$$
$$= \inf\{t: |w(t)| \geq 1\}$$

$\tau(w)$ is the first time the particle hits either of the horizontal lines $+1$ or -1.

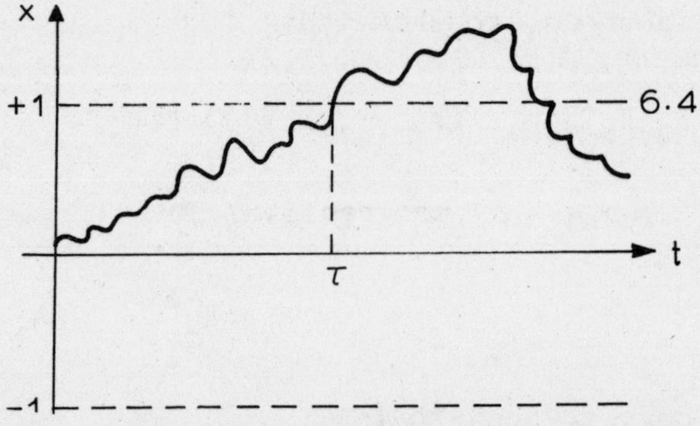

2. Let $\{X_t\}$ be a d-dimensional Brownian motion, G any closed set in $\mathbb{R}^d$. Define

$$\tau(w) = \inf\{t: w(t) \in G\}.$$

This is a generalization of Example 1. To see that τ is a stopping time use

$$\{\tau \leq s\} = \bigcap_{n=1}^{\infty} \overline{\lim_{\substack{\theta \in [0,s] \\ \theta \text{ rational}}}} \{w: w(\theta) \in G_n\},$$

where

$$G_n = \{x \in \mathbb{R}^d : d(x,G) \leq \tfrac{1}{n}\}.$$

3. Let (X_t) be a d-dimensional Brownian motion, C and D disjoint closed sets in $\mathbb{R}^d$. Define

$$\tau(w) = \inf\{t;\ w(t) \in C \text{ and for some } s \leq t, w(s) \in D\}.$$

$\tau(w)$ is the first time that w hits C after visiting D.

5. GENERALISED BROWNIAN MOTION

LET Ω BE ANY space, $\mathcal{F}$ a σ-field and $(\mathcal{F}_t)$ an increasing family of sub σ-fields ssuch that $\sigma(\bigcup \mathcal{F}_t) = \mathcal{F}$. Let P be a measure on $(\Omega, \mathcal{F})$.

$$X(t,w) \colon [0,\infty) \times \Omega \to \mathbb{R}^d$$

is called a *Brownian motion relative to* $(\Omega, \mathcal{F}_t, P)$ if

(i) $X(t,w)$ is progressively measurable with respect to $\mathcal{F}_t$;

(ii) $X(t,w)$ is a.e. continuous in t;

(iii) $X(t,w) - X(s,w)$ for $t > s$ is independent of $\mathcal{F}_s$ and is distributed normally with mean 0 and variance $t-s$, i.e.

$$P(X(t,.) - X(s,.) \in A \mid \mathcal{F}_s) = \int_A \frac{1}{[2\pi(t-s)]^{d/s}} \exp - \frac{|y|^2}{2(t-s)} \, dy.$$

NOTE. 1. The Brownian motion constructed previously was concentrated on $\Omega = C([0,\infty); \mathbb{R}^d)$, $\mathcal{F}$ was the Borel field of Ω, $X(t,w) = w(t)$ and $\mathcal{F}_t = \sigma\{X(s) \colon 0 \leq s \leq t\}$. The measure P so obtained is often called the *Wiener measure*.

2. The above definition is more general because

$$\sigma\{X(s) \colon 0 \leq s \leq t\} \subset \mathcal{F}_t.$$

Exercise. (Brownian motion starting at time s). Let $\Omega = C([s,\infty); \mathbb{R}^d)$, $\mathcal{B}$ = Borel field of Ω. Show that for each $x \in \mathbb{R}^d$ $\exists$ a probability measure P_x^s on Ω such that

(i) $P_x^s \{w: w(s) = x\} = 1$;

(ii) $P_x^s (X_{t_1} \in A_1, \ldots, X_{t_k} \in A_k)$

$$= \int_{A_1} \int_{A_2} \ldots \int_{A_k} p(s,x,t_1,x_1) p(t_1,x_1,t_2,x_2) \ldots p(t_{k-1}, x_{k-1}, t_k, x_k) dx_1 \ldots dx_k,$$

$\forall \; s < t_1 < \ldots < t_k$.

For reasons which will become clear later, we would like to shift the measure P_x^s to a measure on $C([0,\infty); \mathbb{R}^d)$. To do this we define

$$T: C([s,\infty); \mathbb{R}^d) \to C([0,\infty); \mathbb{R}^d)$$

by

$$(Tw)(t) = \begin{cases} w(t), & \text{if } t \geq s, \\ w(s), & \text{if } t \leq s. \end{cases}$$

Clearly, T is continuous. Put

$$P_{s,x} = P_x^s \, T^{-1}.$$

Then

(i) $P_{s,x}$ is a probability measure on the Borel field of $C([0,\infty); \mathbb{R}^d)$;

(ii) $P_{s,x} \{w: w(s) = x\} = 1$.

6. MARKOV PROPERTIES OF BROWNIAN MOTION

NOTATION. 1. A random variable of a stochastic process $\{X(t)\}_{t \in I}$ shall be denoted by X_t or $X(t)$. $0 \leq t < \infty$.

2. $\mathcal{F}_s$ will denote the σ-algebra generated by $\{X_t : 0 \leq t \leq s\}$; $\mathcal{F}_{s+} = \{\mathcal{F}_a : a > s\}$; $\mathcal{F}_{s-}$ will be the σ-algebra generated by $\bigcup \{\mathcal{F}_a : a < s\}$ $s > 0$. It is clear that $\{\mathcal{F}_t\}$ is an increasing family.

3. For the Brownian motion, $\mathcal{B}$ = the σ-algebra generated by $\bigcup \{\mathcal{F}_t : t < \infty\}$ will be denoted by $\mathcal{F}$.

THEOREM. *Let* $\{X_t : 0 \leq t < \infty\}$ *be a Brownian motion. Then* $X_t - X_s$ *is independent of* $\mathcal{F}_s$.

Proof. Let

$$0 \leq t_1 < t_2 < t_3 < \dots < t_k \leq s.$$

Then the σ-algebra generated by $X_{t_1}, \dots, X_{t_k}$ is the same as the σ-algebra generated by

$$X_{t_1}, X_{t_2} - X_{t_1}, \dots, X_{t_k} - X_{t_{k-1}}.$$

Since $X_t - X_s$ is independent of these increments, it is independent of $\sigma\{X_{t_1}, \dots X_{t_k}\}$. This is true for every finite set $t_1, \dots, t_k$ and therefore $X_t - X_s$ is independent of $\mathcal{F}_s$.

Let us carry out the following calculation very *formally*.

$$P[X_t \in A \mid \mathcal{F}_s](w) = P[X_t - X_s \in B \mid \mathcal{F}_s](w), \; B = A - X_s(w),$$

$$= P[X_t - X_s \in B], \text{ by independence,}$$

i.e. $P[X_t \in A | \mathcal{F}_s](w) = \int_A \frac{1}{(2\pi t)^{d/2}} \exp - \frac{|y - X_s(w)|^2}{2(t-s)} dy.$

This formal calculation leads us to

<u>THEOREM</u>. $P[X_t \in A | \mathcal{F}_s](w) = \int_A \frac{1}{(2\pi t)^{d/2}} \exp - \frac{|y-X_s(w)|^2}{2(t-s)} dy.$

where A *is Borel in* $\mathbb{R}^d$, $t > s$.

<u>REMARK.</u> It may be useful to note that $p(s,X_s(w),t,y)$ can be thought of as a conditional probability dens ity.

<u>Proof.</u> (i) We show that

$$f_A(w) = \int_A \frac{1}{(2\pi t)^{d/2}} \exp - \frac{|y-X_s(w)|^2}{2(t-s)} dy$$

is $\mathcal{F}_s$- measurable. Assume first that A is bounded and Borel in $\mathbb{R}^d$. If $\omega_n \to \omega$, then $f_A(\omega_n) \to f_A(\omega)$, i.e. f_A is continuous and hence $\mathcal{F}_s$-measurable. The general case follows if we note that any Borel set can be got as an increasing union of a countable number of bounded Borel sets.

(ii) For any $C \in \mathcal{F}_s$ we show that

$$(*) \qquad \int_C^X X_t^{-1}(A) \, dP(\omega) = \int_C \int_A \frac{\exp-|y-X_s(\omega)|^2}{(2\pi(t-s))^{d/2}} dy \, dP(\omega).$$

It is enough to verify (*) for C of the form

$$C = \{\omega : (X_{t_1}(\omega), \ldots, X_{t_k}(\omega)) \in A_1 \times \ldots \times A_k;\ 0 \le t_1 < \ldots < t_k \le s \quad ,$$

where A_i is Borel in $\mathbb{R}^d$ for $i = 1,2\ldots k$. The left side of (*) is then

$$\int_{A_i \times \dots \times A_k \times A} p(0,0,t_1,x_{t_1})p(t_1,x_{t_1},t_2,x_{t_2})\dots p(t_k,x_{t_k},t,x_t)dx_{t_1}\dots dx_t.$$

To compute the right side define

$$f\colon \mathbb{R}^{(k+1)d} \to \mathbb{R}$$

by

$$f(u_1,\dots u_k,u) = \chi_{A_1}(u_1)\dots\chi_{A_k}(u_k)p(s,u,t,y).$$

Clearly f is Borel measurable. An application of Fubini's theorem to the right side of (*) yields

$$\int_A dy \int_\Omega \chi_{A_1}(X_{t_1}(\omega))\dots\chi_{A_k}(X_{t_k}(\omega))p(s,X_s(\omega),t,y)dP(\omega)$$

$$= \int_A dy \int_{\underbrace{\mathbb{R}^d \times \dots \times \mathbb{R}^d}_{(k+1)\text{ times}}} f(x_1\dots x_k,x_s)d\,F_{t_1\dots t_k,s}$$

$$= \int_A dy \int_{A_1\times\dots\times A_k\times \mathbb{R}^d} p(0,0,t_1,x_1)\dots p(t_{k-1},t_k,x_k)$$

$$p(t_k,x_k,s,x_s)p(s,x_s,t,y)dx_1\dots dx_k\, dx_s$$

$$= \int_{A_1\times\dots\times A_k\times A} p(0,0,t_1,x_1)\dots p(t_{k-1},x_{k-1},t_k,x_k)p(t_k,x_k,t,y)\, dx_1\dots dx_k\, dy$$

(b y the convolution rule)

= left side.

EXAMPLES OF STOPPING TIMES.

1. Let (X_t) be a one-dimensional Brownian motion. Define τ by

$$\{\tau \leq s\} = \bigcap_{n=1}^{\infty} \overline{\lim_{\substack{\theta_1,\theta_2 \\ \theta_1,\theta_2 \text{ rational in } [0,s]}}} \{w: w(\theta_1) \in D_n, w(\theta_2) \in C_n\},$$

where

$$D_n = \{x \in \mathbb{R}^d: d(x,D) \leq \tfrac{1}{n}\}, C_n = \{x \in \mathbb{R}^d: d(x,C) \leq \tfrac{1}{n}\}$$

Exercise 1. Let τ be as in Example 1.

(a) If $A = \{w: X_1(w) \leq \tau\}$ show that $A \notin \mathcal{F}_\tau$.

(Hint: $A \cap \{\tau \leq 0\} \notin \mathcal{F}_o$). This shows that $\mathcal{F}_\tau \subsetneq \mathcal{F}_o$.

(b) $$P_o \{w: \tau(w) = \infty\} = 0.$$

(Hint: $P_o \{w: |w(t)| < 1\} \leq \int_{|y| \leq t^{-1/2}} e^{-1/2|y|^2} dy \;\; \forall\; t$).

THEOREM.- *(Strong Markov Property of Brownian Motion). Let* τ *be any finite stopping time, i.e.* $\tau < \infty$ *a.e. Let* $Y_t = X_{\tau+t} - X_\tau$. *Then*

1. $P[(Y_{t_1} \in A_1,\ldots,Y_{t_k} \in A_k) \cap A] = P(X_{t_1} \in A_1,\ldots X_{t_k} \in A_k).P(A)$, $\forall\; A \in \mathcal{F}_\tau$ *and for every* A_i *Borel in* $\mathbb{R}^d$. *Consequently,*
2. (Y_t) *is a Brownian motion.*
3. (Y_t) *is independent of* $\mathcal{F}_\tau$.

The assertion is that a Brownian motion starts afresh at every stopping time.

Proof. Step 1. Let τ take only countably many values, say $s_1, s_2, s_3 \ldots$. Put $E_j = \tau^{-1}\{s_j\}$. Then each E_j is $\mathcal{F}_\tau$-measurable and

$$\Omega = \bigcup_{j=1}^{\infty} E_j,\ E_j \cap E_i = \emptyset \quad j \neq i.$$

Fix $A \in \mathcal{F}_\tau$.

$$P[(Y_{t_1} \in A_1,\ldots,Y_{t_k} \in A_k) \cap A]$$

$$= \sum_{j=1}^{\infty} P[(Y_{t_1} \in A_1,\ldots,Y_{t_k} \in A_k) \cap A \cap E_j]$$

$$= \sum_{j=1}^{\infty} P[(X_{t_1+s_j} - X_{s_j}) \in A_1,\ldots,X_{t_k+s_j} - X_{s_j} \in A_k) \cap A \cap E_j]$$

$$= \sum_{j=1}^{\infty} P[(X_{t_1} \in A_1),\ldots,(X_{t_k} \in A_k)]P(A \cap E_j) \text{(by the Markov property)}$$

$$= P(X_{t_1} \in A_1,\ldots,X_{t_k} \in A_k).\ P(A)$$

<u>Step 2.</u> Let τ be any stopping time; put $\tau_n = \frac{[n\tau]+1}{n}$. A simple calculation shows that τ_n is a stopping time taking only countably many values. As $\tau_n \downarrow \tau$, $\mathcal{F}_\tau \subset \mathcal{F}_{\tau_n}$ $\forall_n$. Let $Y_t^{(n)} = X_{\tau_n+t} - X_{\tau_n}$.

By Step 1,

$$P[(Y_{t_1}^{(n)} < x_1,\ldots,Y_{t_k}^{(n)} < x_k) \cap A]$$

$$= P(X_{t_1} < x_1,\ldots,X_{t_k} < x_k).\ P(A)$$

(where $x < y$ means $x_i < y_i$ $i = 1,2,\ldots,d$) for every $A \in \mathcal{F}_\tau$. As all the Brownian paths are continuous, $Y_t^{(n)} \to Y_t$ a.e. Thus, if $x_1,\ldots x_k$ is a point of continuity of the joint distribution of $X_{t_1},\ldots,X_{t_k}$, we have

$$P[(Y_{t_1} < x_1,\ldots,Y_{t_k} < x_k) \cap A] = P(X_{t_1} < x_1,\ldots,X_{t_k} < x_k)\, P(A)$$

$\forall\ A \in \mathcal{F}_\tau$. Now assertion (1) follows easily.

For (2), put $A = \Omega$ in (1), and (3) is a consequence of (1) and (2).

7. REFLECTION PRINCIPLE

LET (X_t) BE A one-dimensional Brownian motion. Then $P(\sup_{0\leq s\leq t} X_s \geq a) = 2P(X_t \geq a)$ with $a > 0$. This gives us the probability of a Brownian particle hitting the line $x = a$ some time less than or equal to t. The intuitive idea of the proof is as follows.

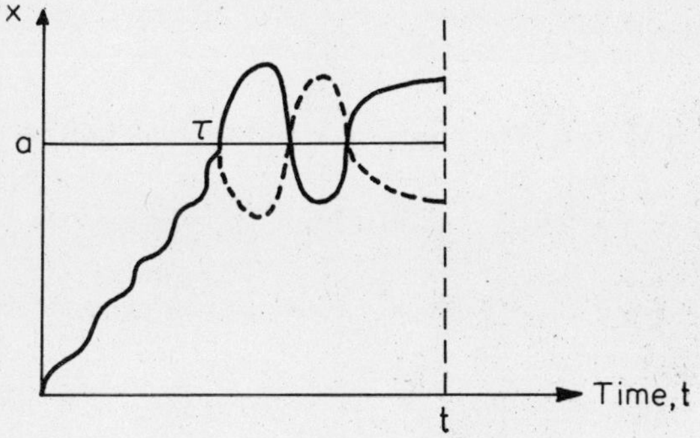

Among all the paths that hit a before time t exactly half of them end up below a at time t. This is due to the reflection symmetry. If $X_s = a$ for some $s < t$, reflection about the horizontal line at a gives a one - one correspondence between paths with $X_t > a$ and paths with $X_t < a$. Therefore

$$P\{\max_{0\leq s\leq t} X_s \geq a, X_t > a\} = P\{\max_{0\leq s\leq t} X_s \geq a, X_t < a\}$$

Since $P\{X_t = a\} = 0$, we obtain

$$P\{\sup_{0\leq s\leq t} X_s \geq a\} = P\{\sup_{0\leq s\leq t} X_s \geq a, X_t > a\} + P\{\sup_{0\leq s\leq t} X_s \geq a, X_t > a$$

$$= 2P\{X_t \geq a\}$$

We shall now give a precise argument. We need a few elementary results.

LEMMA 1. *Let* $X_n = \sum_{k=1}^{n} Y_k$ *where the* Y_k *are independent random variables such that* $P\{Y_k \in B\} = P\{-Y_k \in B\}$ $\forall$ *Borel set* $B \subset R$ *(i.e.* Y_k *are symmetric). Then for any real number* a,

$$P\{\max_{1 \leq i \leq n} X_i > a\} \leq 2P\{X_n > a\}$$

Proof. It is easy to verify that a random variable is symmetric if and only if its characteristic function is real. Define

$$A_i = \{X_1 \leq a, .. X_{i-1} \leq a, X_i > a\},\ i = 1,2,...,n;$$

$$B = \{X_n > a\}$$

Then $A_i \cap A_j = \emptyset$ if $i \neq j$. Now,

$$\begin{aligned} P(A_i \cap B) &\geq P(A_i \cap \{X_n \geq X_i\}) \\ &= P(A_i)P(X_n \geq X_i), \text{ by independence.} \\ &= P(A_i)P(Y_{i+1} + ... + Y_n \geq 0). \end{aligned}$$

As $Y_{i+1}, .., Y_n$ are independent, the characteristic function of $Y_{i+1} + ... + Y_n$ is the product of the characteristic functions of $Y_{i+1} + ... + Y_n$, so that $Y_{i+1} + ... + Y_n$ is symmetric. Therefore

$$P(Y_{i+1} + ... + Y_n \geq 0) \geq \frac{1}{2}.$$

Thus $P(A_i \cap B) \geq \frac{1}{2} P(A_i)$ and

$$P(B) \geq \sum_{i=1}^{n} P(A_i \cap B) \geq \frac{1}{2}\sum P(A_i) \geq \frac{1}{2}P(\bigcup_{i=1}^{n} A_i),$$

i.e. $$2P(B) \geq P\left(\bigcup_{i=1}^{n} A_i\right),$$

or

$$P\{\max_{1\leq i\leq n} X_i > a\} \leq 2P\{X_n > a\}$$

<u>LEMMA 2</u>. *Let* $Y_1,\ldots,Y_n$ *be independent random variables. Put* $X_n = \sum_{k=1}^{n} Y_k$ *and let* $\tau = \min\{i: X_i > a\}$, $a > 0$ *and* $\tau = \infty$ *if there is no such* i. *Then for each* $\varepsilon > 0$,

(a) $P\{\tau \leq n-1, X_n - X_\tau \leq -\varepsilon\} \leq P\{\tau \leq n-1, X_n \leq a\} + \sum_{j=1}^{n-1} P(Y_j > \varepsilon)$.

(b) $P\{\tau \leq n-1, X_n > a+2\varepsilon\} \leq P\{\tau \leq n-1, X_n - X_\tau > \varepsilon\} + \sum_{j=1}^{n-1} P\{Y_j > \varepsilon\}$

(c) $P\{X_n > a+2\varepsilon\} \leq P\{\tau \leq n-1, X_n > a+2\varepsilon\} + P\{Y_n > 2\varepsilon\}$.

If, further, $Y_1,\ldots,Y_n$ *are symmetric, then*

(d) $P\{\max_{1\leq i\leq n} X_i > a, X_n \leq a\} \geq P\{X_n > a+2\varepsilon\} - P\{Y_n \geq 2\varepsilon\} - 2\sum_{j=1}^{n-1} P\{Y_j > \varepsilon\}$

(e) $P\{\max_{1\leq i\leq n} X_i > a\} \geq 2P\{X_n > a+2\varepsilon\} - 2\sum_{j=1}^{n} P\{Y_j > \varepsilon\}$

<u>Proof (a)</u>. Suppose $w \in \{\tau \leq n-1, X_n - X_\tau \leq -\varepsilon\}$ and $w \notin \{\tau \leq n-1, X_n \leq a\}$. Then $X_n(w) > a$ and $X_n(w)+\varepsilon \leq X_{\tau(w)}(w)$ or, $X_{\tau(w)}(w) > a + \varepsilon$.

By definition of $\tau(w)$, $X_{\tau(w)-1}(w) \leq a$ and therefore,

$$Y_{\tau(w)}(w) = X_{\tau(w)}(w) - X_{\tau(w)-1}(w) > a + \varepsilon - a = \varepsilon$$

if $\tau(w) > 1$; if $\tau(w) = 1$, $Y_{\tau(w)}(w) = X_{\tau(w)}(w) > a+\varepsilon > \varepsilon$.

Thus $Y_j(w) > \varepsilon$ for some $j \leq n-1$,

i.e. $$w \in \bigcup_{j=1}^{n-1} \{Y_j > \varepsilon\}.$$

Therefore

$$\{\tau \leq n-1, X_n - X_\tau \leq -\varepsilon\} \subset \{\tau \leq n-1, X_n \leq a\} \bigcup_{j=1}^{n-1} \{Y_j > \varepsilon\}$$

and (a) follows.

(b) Suppose $w \in \{\tau \leq n-1, X_n > a+2\varepsilon\}$ but $w \in \{\tau \leq n-1, X_n - X_\tau > \varepsilon\}$.

Then

$$X_n(w) - X_{\tau(w)}(w) \leq \varepsilon,$$

or, $X_{\tau(w)}(w) > a+\varepsilon$ so that $Y_{\tau(w)}(w) > \varepsilon$ as in (a); hence $Y_j(w) > \varepsilon$ for some $j \leq n-1$. This proves (b).

(c) If $w \in \{X_n > a+2\varepsilon\}$, then $\tau(w) \leq n$; if $w \notin \{\tau \leq n-1, X_n > a+2\varepsilon\}$, then $\tau(w) = n$ so that $X_{n-1}(w) \leq a$; therefore $Y_n(w) = X_n(w) - X_{n-1}(w) > 2\varepsilon$. i.e. $w \in \{Y_n > 2\varepsilon\}$.

Therefore

$$\{X_n > a+2\varepsilon\} \subset \{\tau \leq n-1, X_n > a+2\varepsilon\} \cup \{Y_n > 2\varepsilon\}.$$

This establishes (c).

(d) $$P\{\max_{1 \leq i \leq n} X_i > a, X_n \leq a\} = P\{\tau \leq n-1, X_n \leq a\}$$

$$\geq P\{\tau \leq n-1, X_n - X_\tau \leq -\varepsilon\} - \sum_{j=1}^{n-1} P(Y_j > \varepsilon), \text{ by (a)},$$

$$= P\left[\bigcup_{k=1}^{n-1} \{\tau = k, X_n - X_k \leq -\varepsilon\}\right] - \sum_{j=1}^{n-1} P(Y_j > \varepsilon)$$

$$= \sum_{k=1}^{n-1} P\{\tau = k, X_n - X_k \leq -\varepsilon\} - \sum_{j=1}^{n-1} P(Y_j > \varepsilon)$$

$$= \sum_{k=1}^{n-1} P\{\tau = k\} \, P\{X_n - X_k \leq -\varepsilon\} - \sum_{j=1}^{n-1} P(Y_j > \varepsilon) \quad \text{(by independence)}$$

$$= \sum_{k=1}^{n} P\{\tau = k\} \, P\{X_n - X_k > \varepsilon\} - \sum_{j=1}^{n-1} P(Y_j > \varepsilon) \quad \text{(by symmetry)}$$

$$= P\{\tau \leq n-1, \; X_n - X_\tau \geq \varepsilon\} - \sum_{j=1}^{n-1} P(Y_j > \varepsilon)$$

$$\geq P\{\tau \leq n-1, \; X_n - X_\tau > \varepsilon\} - \sum_{j=1}^{n-1} P(Y_j > \varepsilon)$$

$$\geq P\{\tau \leq n-1, X_n > a+2\varepsilon\} - 2\sum_{j=1}^{n-1} P\{Y_j > \varepsilon\} \quad \text{(by (b))}$$

$$\geq P\{X_n > a+2\varepsilon\} - P\{Y_n > 2\varepsilon\} - 2\sum_{j=1}^{n-1} P\{Y_j > \varepsilon\} \quad \text{(by (c))}$$

This proves (d).

(e) $$P\{\max_{1 \leq i \leq n} X_i > a\} = P\{\max_{1 \leq i \leq n} X_i > a, X_n \leq a\} + P\{\max_{1 \leq i \leq n} X_i > a, X_n > a\}$$

$$= P\{\max_{1 \leq i \leq n} X_i > a, \; X_n \leq a\} + P\{X_n > a\}$$

$$\geq P\{X_n > a+2\varepsilon\} - P\{Y_n > 2\varepsilon\} + P\{X_n > a\}$$

$$-2\sum_{j=1}^{n-1} P\{Y_j > \varepsilon\} \quad \text{(by (d))}$$

Since $P\{X_n > a+2\varepsilon\} \leq P\{X_n > a\}$ and

$$P\{Y_n > 2\varepsilon\} \leq P\{Y_n > \varepsilon\} \leq 2P\{Y_n > \varepsilon\},$$

we get

$$P\{\max_{1\le i\le n} X_i > a\} \ge 2P\{X_n > a+2\varepsilon\} - 2\sum_{j=1}^{n} P(Y_j > \varepsilon)$$

This completes the proof.

Proof of the reflection principle.

By Lemma 1,

$$p = P\{\max_{1\le j\le n} X(\tfrac{jt}{n}) > a\} \le 2P(X(t) > a).$$

By Lemma 2(e),

$$p \ge 2P(X(t) > a+2\varepsilon) - 2\sum_{j=1}^{n} P\{(X(\tfrac{jt}{n}) - X(\tfrac{(j-1)t}{n})) > \varepsilon\}.$$

Since $X(\frac{jt}{n}) - X(\frac{(j-1)t}{n})$ are independent and identically distributed normal random variables with mean zero and variance $\frac{t}{n}$ (in particular they are symmetric),

$$P((X(\tfrac{jt}{n}) - X(\tfrac{(j-1)t}{n})) > \varepsilon) = P((X(\tfrac{t}{n}) - X(0)) > \varepsilon)$$

$$= P(X(\tfrac{t}{n}) > \varepsilon).$$

Therefore

$$p \ge 2P(X(t) > a+2\varepsilon) - 2n\, P(X(\tfrac{t}{n}) > \varepsilon).$$

$$P(X(t/n) > \varepsilon) = \int_{\varepsilon}^{\infty} \frac{1}{\sqrt{(2\,t/n)}} e^{-x^2/\frac{2t}{n}}\, dx$$

$$= \int_{\varepsilon\sqrt{n}/\sqrt{(2t)}}^{\infty} \frac{e^{-x^2}}{\sqrt{\pi}}\, dx \le \frac{\varepsilon\sqrt{n}/\sqrt{(2t)}}{\sqrt{\pi}}^{-1} \int_{\varepsilon\sqrt{n}/\sqrt{(2t)}}^{\infty} x e^{-x^2}\, dx$$

or

$$P(X(t/n) > \varepsilon) \le \frac{1}{2\sqrt{\pi}} e^{-\varepsilon^2 n/2t} \cdot \frac{\sqrt{(2t)}}{\varepsilon\sqrt{n}}\,.$$

Therefore

$$nP(X(t/n) > \varepsilon) \le \frac{n}{2\varepsilon} \surd(2t)/\surd(\pi n)\ e^{-\varepsilon^2 n/2t} \to 0 \quad \text{as} \quad n \to +\infty.$$

By continuity,

$$P\{\max_{1 \le j \le n} X(jt/n) > a\} \to P\{\max_{0 \le s \le t} X(s) > a\} .$$

We let n tend to ∞ through values $2, 2^2, 2^3, \ldots.$ so that we get

$$2P\{X(t) > a+2\varepsilon\} - 2n\, P\{X(t/n) > \varepsilon\}$$

$$\le P\{\max_{1 \le j \le n} X(t/n) > a\} \le 2P\{X(t) > a\} ,$$

or

$$2P\{X(t) > a\} \le 2P\{X(t) \ge a\} \le P\{\max_{0 \le s \le t} X(s) > a\}$$

$$2P\{X(t) > a\} ,$$

on letting $n \to +\infty$ first and then letting $\varepsilon \to 0$. Therefore,

$$P\{\max_{0 \le s \le t} X(s) > a\} = 2P\{X(t) > a\}$$

$$= 2 \int_a^\infty 1/\surd(2\pi t)\ e^{-x^2/2t}\, dx.$$

<u>AN APPLICATION.</u> Consider a one-dimensional Brownian motion. A particle starts at 0. What can we say about the behaviour of the particle in a small interval of time $[0,\varepsilon)$? The answ er is given by the following result.

$$P(A) \equiv P\{w: \forall\ \varepsilon > 0, \exists\ t,s \text{ in } [0,\varepsilon) \text{ such that } X_t(w) > 0 \text{ and }$$

$$X_s(w) < 0\} = 1.$$

INTERPRETATION. Near zero all the particles oscillate about their starting point. Let

$$A^+ = \{w: \forall\ \varepsilon > 0\ \exists\ t \in [0,\varepsilon) \text{ such that } X_t(w) > 0\},$$

$$A^- = \{w: \forall\ \varepsilon > 0\ \exists\ s \in [0,\varepsilon) \text{ such that } X_s(w) < 0\}.$$

We show that $P(A^+) = P(A^-) = 1$ and therefore $P(A) = P(A^+ \cap A^-) = 1$.

$$A^+ \supset \bigcap_{n=1}^{\infty} \{ \sup_{0 \le t \le 1/n} w > 0\} = \bigcap_{n=1}^{\infty} \bigcup_{m=1}^{\infty} (\sup_{0 \le t \le 1/n} w(t) \ge 1/m)$$

Therefore

$$P(A^+) \ge \operatorname*{Lt}_{n\to\infty} \sup_{m\to\infty} P(\sup_{0 \le t \le 1/n} w(t) \ge 1/m)$$

$$\ge 2 \operatorname*{Lt}_{n\to\infty} \sup_{m\to\infty} P(w(1/n) \ge 1/m) \quad \text{(by the reflection principle)}$$

$$\ge 1.$$

Similarly $P(A^-) = 1$.

THEOREM. *Let* $\{X_t\}$ *be a one-dimensional Brownian motion,* $A \subset (-\infty,a)$ $(a > 0)$ *any Borel subset of* $\mathbb{R}$. *Then*

$$P_o\{X_t \in A, X_s < a\ \forall\ s \text{ such that } 0 \le s \le t\}$$

$$= \int_A 1/\sqrt{(2\pi t)}\, e^{-y^2/2t}\, dy - \int_A 1/\sqrt{(2\pi t)}\, e^{-(2a-y)^2/2t}\, dy$$

Proof. Let $\tau(w) = \inf \{t: w(t) \ge a\}$. By the strong Markov property of Brownian motion,

$$P_o\{B \quad (X(\tau+s) - X(\tau) \in A)\} = P_o(B)\, P_o(X(s) \in A)$$

for every set B in $\mathcal{F}_t$. This can be written as

$$E(\chi_{(X(\tau+s)-X(\tau) \in A)} | \mathcal{F}_\tau) = P_o(X(s) \in A)$$

Therefore

$$E(\chi_{(X(\tau+\ell(w))-X(\tau) \in A)} | \mathcal{F}_\tau) = P_o(X(\ell(w\ )) \in A)$$

for every function $\ell(w)$ which is $\mathcal{F}_\tau$-measurable. Therefore,

$$P_o((\tau \leq t) \cap ((X(\tau+\ell(w))-X(\tau)) \in A) = \int_{\{\tau \leq t\}} P_o(X(\ell(w)\ ) \in A)dP(w)$$

In particular, take $\ell(w) = t - \tau(w)$, clearly $\ell(w)$ is $\mathcal{F}_\tau$-measurable. Therefore,

$$P_o((\tau \leq t) \quad ((X(t)-X(\tau)) \in A)) = \int_{\{\tau \leq t\}} P_o(X(\ell(w) \in A)\ dP(w).$$

Now $X(\tau(w)) = a$. Replace A by $A-a$ to get

$$(*) \quad P_o((\tau \leq t) \cap (X(t) \in A)) = \int_{\{\tau \leq t\}} P_o(X(\ell(w) \in A-a)\ dP(w)$$

Consider now

$$\begin{aligned} P_{2a}(X(t) \in A) &= P_o(X(t) \in A-2a) \\ &= P_o(X(t) \in 2a-A) \quad \text{(by sy mmetry of } x) \\ &= P_o((\tau \leq t) \cap (X(t) \in 2a-A)). \end{aligned}$$

The last step follows from the fact that $A \subset (-\infty,a)$ and the continuity of the Brownian paths. Therefore

$$\begin{aligned} P_{2a}(X(t) \in A) &= \int_{\{\tau \leq t\}} P_o(X(\ell(w)) \in a-A)\ dP(w), \text{ (using *)} \\ &= P_o((\tau \leq t) \cap (X(t) \in A)). \end{aligned}$$

Now the required probability

$$P_o\{X_t \in A, X_s < a\ \forall\ s \ni 0 \leq s \leq t\} = P_o\{X_t \in A\} - P_o\{(\tau \leq t) \cap (X_t \in A)\}$$

$$= \int_A 1/\sqrt{(2\pi t)}\ e^{-y^2/2t}\ dy - \int_A 1/\sqrt{(2\pi t)}\ e^{-(2a-y)^2/2t} dy.$$

The intuitive idea of the previous theorem is quite clear. To obtain the paths that reach A at time t without hitting the horizontal line $x = a$, we consider all paths that reach A at time t and subtract those paths that hit the horizontal line $x = a$ before time t and then reach A at time t. To see exactly which paths reach A at time t after hitting $x = a$ we consider a typical path $X(w)$.

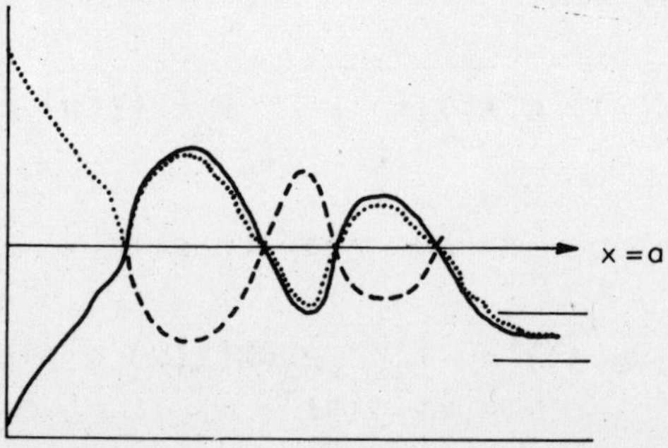

The reflection principle (or the strong Markov property) allows us to replace this path by the dotted path (see Fig.). The symmetry of the Brownian motion can then be used to reflect this path about the line $x = a$ and obtain the path shown in dark. Thus we have the following result:

the probability that a Brownian particle starts from $x = 0$ at $t = 0$ and reaches A at time t after it has hit $x = a$ at some time $\tau \leq t$ is the same as if the particle started at time $t = 0$ at $x = 2a$ and reached A at time t. (The continuity of the path ensures that at some time $\tau \leq t$, this particle has to hit $x=a$).

We shall use the intuitive approach in what follows, the mathematical analy sis being clear, though lengthy.

<u>THEOREM</u>. *Let* $X(t)$ *be a one-dimensional Brownian motion,* $A \subset (-1,1)$ *any Borel subset of* $\mathbb{R}$. *Then*

$$P_o[\sup_{0 \le s \le t} |X(s)| < 1,\ X(t) \in A] = \int_A \phi(t,y)\, dy,$$

where

$$\phi(t,y) = \sum_{n=-\infty}^{\infty} (-1)^n/\sqrt{(2\pi t)}\ e^{-(y-2n)^2/2t}.$$

<u>Proof</u>.

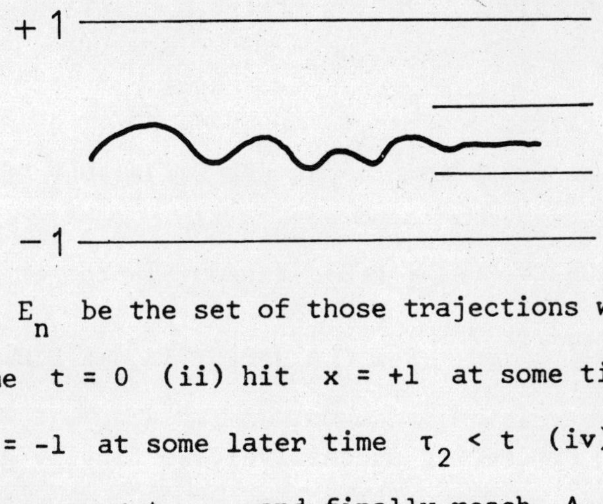

Let E_n be the set of those trajections which (i) start at $x = 0$ at time $t = 0$ (ii) hit $x = +1$ at some time $\tau_1 < t$ (iii) hit $x = -1$ at some later time $\tau_2 < t$ (iv) hit $x = 1$ again at a later time $\tau_3 < t$ ... and finally reach A at time t. The number of τ's should be equal to n at least, i.e.

E_n = {w: there exists a sequence $\tau_1, \dots \tau_n$ of stopping times such that $0 < \tau_1 < \tau_2 < \dots < \tau_n < t, X(\tau_j) = (-1)^{j-1}, X(t) \in A$}.

Similarly, let

F_n = {w: there exists a s equence $\tau_1, \dots, \tau_n$ of stopping times $0 < \tau_1 < \tau_2 < \dots < \tau_n < t$, $X(\tau_j) = (-1)^j, X(t) \in A$}

Note that $E_1 \supset E_2 \supset E_3 \supset \ldots,$

$F_1 \supset F_2 \supset F_3 \supset \ldots,$

$F_n \supset E_{n+1}\,;\; E_n \supset F_{n+1},$

$E_n \cap F_n = E_{n+1} \cup F_{n+1}.$

Let

$$\phi(t,A) = P[\sup_{0 \leq s \leq t} |X(s)| < 1,\ X(t) \in A].$$

Therefore

$$\phi(t,A) = P[X(t) \in A] - P[\sup_{0 \leq s \leq t} |X(s)| \geq 1, X(t) \in A]$$

$$= \int_A 1/\sqrt{(2\pi t)}\, e^{-y^2/2t}\, dy - P[(E_1 \cup F_1) \cap A_o],$$

where

$$A_o = \{X(t) \in A\} = \int_A 1/\sqrt{(2\pi t)}\, e^{-y^2/2t} dy - P[(E_1 \cap A_o) \cup (F_1 \cap A_o)].$$

Use the fact that $P[A \cup B] = P(A) + P(B) - P(A \cap B)$ to get

$$\phi(t,A) = \int_A 1/\sqrt{(2\pi t)} e^{-y^2/2t} dy \quad - P[E_1 \cap A_o] - P[F_1 \cap A_o] + P[E_1 \cap F_1 \cap A_o],$$

as $E_1 \cap F_1 = E_2 \cup F_2$. Proceeding successively we finally get

$$\phi(t,A) = \int_A 1/\sqrt{(2\pi t)} e^{-y^2/2t} dy + \sum_{n=1}^{\infty} (-1)^n P[E_n \cap A_o] + \sum_{n=1}^{\infty} (-1)^n P[F_n \cap A_o]$$

We shall obtain the express ion for $P(E_1 \cap A_o)$ and $P[E_2 \cap A_o]$, the other terms can be obtained similarly.

$E_1 \cap A_o$ consis ts of those traject ries that hit $x = +1$ at some time $\tau \leq t$ and then reach A at time t. Thus $P[E_1 \cap A_o]$ is given by the previous theorem by

$$\int_A 1/\sqrt{(2\pi t)}\, e^{-(y-2)^2/2t}\, dy.$$

$E_2 \cap A_o$ consists of those trajectories that hit $x = +1$ at time τ_1, hit $x = -1$ at time τ_2 and reach A at time t $(\tau_1 < \tau_2 < t)$.

According to the previous theorem we can reflect the trajectory upto τ_2 about $x = -1$ so that $P(E_2 \cap A_o)$ is the same as if the particle starts at $x = -2$ at time $t = 0$, hits $x = -3$ at time τ_1 and ends up in A at time t. We can now reflect the trajectory

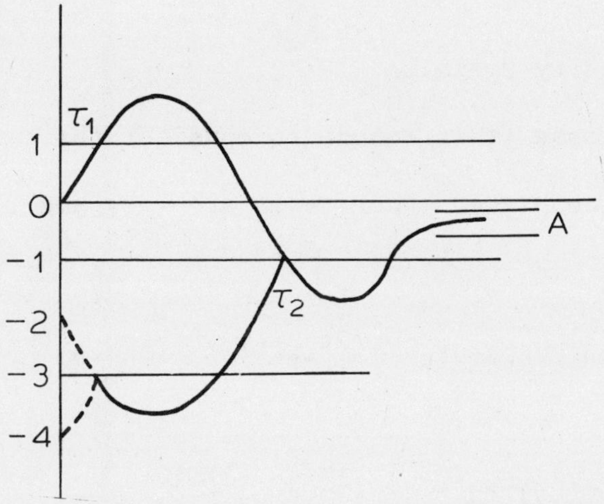

upto time τ_1 (the dotted curve should be reflected) about $x = -3$ to obtain the required probability as if the trajectory started at $x = -4$. Thus,

$$P(E_2 \cap A_o) = \int_A e^{-(y+4)^2/2t}/\sqrt{(2\pi t)}\, dy.$$

Thus

$$\phi(t,A) = \sum_{n=-\infty}^{\infty} (-1)^n \int_A \frac{e^{-(y-2n)^2/2t}}{1/\sqrt{(2\pi t)}}\, dy$$

$$= \int_A \phi(t,y)\, dy.$$

The previous theorem leads to an interesting result:

$$P[\sup_{0 \le s \le t} |X(s)| < 1] = \int_{-1}^{1} \phi(t,y)\, dy$$

Therefore

$$P[\sup_{0\le s\le t} |X(s)| \ge 1] = 1 - P[\sup_{0\le s\le t} |X(s)| < 1]$$

$$= -1 - \int_{-1}^{1} \phi(t,y)\, dy ,$$

$$\phi(t,y) = \sum_{n=-\infty}^{\infty} (-1)^n/\surd(2\pi t)\; e^{-(y-2n)^2/2t}$$

<u>Case (i)</u>. t is very small.

In this case it is enough to consider the terms corresponding to $n = 0, \pm 1$ (the higher order terms are very small). As y varies from -1 to 1,

$$\phi(t,y) \simeq 1/\surd(2\pi t)[e^{-y^2/2t} - e^{-(y-2)^2/2t} - e^{-(y+2)^2/2t}] .$$

Therefore

$$\int_{-1}^{1} \phi(t,y)\, dy \simeq 4/\surd(2\pi t)\; e^{-1/2t}.$$

<u>Case (ii)</u>. t is large. In this cas e we use Poisson's summation formula for $\phi(t,y)$:

$$\phi(t,y) = \sum_{k=0}^{\infty} e^{-(2k+1)^2\pi^2 t/8} \operatorname{Cos}\{(k+1)/2\ \pi y\} ,$$

to get

$$\int_{-1}^{1} \phi(t,y)\, dy \simeq 4/\pi\; e^{-\pi^2 t/8}$$

for large t. Thus, $P(\tau > t) \approx 4/\pi \, e^{-\pi^2 t/8}$

This result says that for large values of t the probability of paths which stay between -1 and $+1$ is very very small and the decay rate is governed by the factor $e^{-\pi^2 t/8}$. This is connected with the solution of a certain differential equation as shall be seen later on.

8. BLUMENTHAL'S ZERO-ONE LAW

LET X_t BE A d-dimensional Brownian motion. If $A \in \mathcal{F}_{o+} = \bigcap_{t>0} \mathcal{F}_t$, then $P(A) = 0$ or $P(A) = 1$.

Interpretation. If an event is observable in every interval $[0,t]$ of time then either it always happens or it never happens.

We shall need the following two lemmas.

LEMMA 1. *Let* $(\Omega, \mathcal{B}, P)$ *be any probability space,* $\mathcal{C}$ *a sub-σ-algebra of* $\mathcal{B}$. *Then*

(a) $L^2(\Omega, \mathcal{C}, P)$ *is a closed subspace of* $L^2(\Omega, \mathcal{B}, P)$.

(b) If $\Pi: L^2(\Omega, \mathcal{B}, P) \to L^2(\Omega, \mathcal{C}, P)$ *is the projection map then* $\Pi f = E(f|\mathcal{C})$.

Proof. Refer appendix.

LEMMA 2. *Let* $\Omega = C([0,\infty); \mathbb{R}^d), P_o$ *the probabi lity corresponding to the Brownian motion. Then the set* $\{\phi(\Pi_{t_1}, \ldots, t_k) \ni \phi$ *is continuous, bounded on* $\mathbb{R}^d \times \ldots \times \mathbb{R}^d$ *(k times),* $\Pi_{t_1}, \ldots, t_k$ *the canonical projection} is dense in* $L^2(\Omega, \mathcal{B}, P)$.

Proof. Functions of the form $\phi(x(t_1), \ldots, x(t_k))$ where ϕ runs over continuous functions is clearly dens e in $L_2(\Omega, \mathcal{F}_{t_1, t_2, \ldots, t_k}, P)$

and $\bigcup_k \bigcup_{t_1, \ldots, t_k} L_2(\Omega, \mathcal{F}_{t_1, \ldots, t_k}, P)$ is clearly dense in $L_2(\Omega, \mathcal{B}, P)$.

Proof of zero-one law. Let

$$H_t = L^2(\Omega, \mathcal{F}_t, P),\ H = L^2(\Omega, \mathcal{B}, P),\ H_{o+} = \bigcap_{t > 0} H_t.$$

Clearly $H_{o+} = L^2(\Omega, \mathcal{F}_{o+}, P)$.

Let $\Pi_t : H \to H_t$ be the projection. Then $\Pi_t f \to \Pi_{o+} f \ \forall\ f$ in H. To prove the law it is enough to show that H_{o+} contains only constants, which is equivalent to $\Pi_{o+} f$ = constant $\forall\ f$ in H. As Π_{o+} is continuous and linear it is enough to show that

$\Pi_{o+}\ \phi$ = const $\forall\ \phi$ of the Lemma 2:

$$\Pi_{o+}\ \phi = \underset{t\to 0}{\text{Lt}}\ \Pi_t\ \phi = \underset{t\to 0}{\text{Lt}}\ E(\phi|\ \ _t) \text{ by Lemma 1,}$$

$$= \underset{t\to 0}{\text{Lt}}\ E(\phi(t_1,..t_k)|\mathcal{F}_t).$$

We can assume without loss of generality that $t < t_1 < t_2 <\ldots< t_k$.

$$E(\phi(t_1,\ldots,t_k)|\mathcal{F}_t) = \int \phi(y_1,\ldots,y_k) 1/\sqrt{(2\pi(t_1-t))}\, e^{-|y_1-X_t(w)|^2/2(t_1-t)} \ldots$$

$$\ldots\ 1/\sqrt{(2\pi(t_k-t_{k-1}))}\, e^{-|y_k-y_{k-1}|^2/2(t_k-t_{k-1})} dy_1..dy_k.$$

Since $X_o(w) = 0$ we get, as $t \to 0$,

$$\Pi_{o+}\ \phi = \text{cons tant.}$$

This completes the proof.

APPLICATION. Let $\alpha \geq 1$ $A = \{w: \int_0^1 |w(t)|/t^\alpha < \infty\}$. Then $A \in \mathcal{F}_{o+}$.

For, if $0 < s < 1$, then $\int_s^1 |w(t)|/t^\alpha < \infty$. Therefore $w \in A$ or not according as $\int_o^s |w(t)|/t^\alpha\, dt$ converges or not. But this convergence can be asserted by knowing the history of w upto time s. Hence $A \in \mathcal{F}_s$. Blumenthal's law implies that

$$\int_0^1 |w(t)|/t^\alpha \, dt < \infty \quad \text{a.e.w., or,} \quad \int_0^1 |w(t)|/t^\alpha \, dt = \infty \quad \text{a.e.w.}$$

A precise argument can be given along the following lines. If $0 < s < 1$,

$$A = \{w: \int_0^s |w(t)|/t^\alpha < \infty\}$$

$$= \{w: \sup I_{n,s}(w) < \infty\}$$

where $I_{n,s}(w)$ is the lower Riemannian sum of $|w(t)^n|/t^\alpha$ corresponding to the partition $\{0, s/n, \ldots, s\}$ and each $I_{n,s} \in \mathcal{F}_s$.

9. PROPERTIES OF BROWNIAN MOTION IN ONE DIMENSION

WE NOW PROVE the following.

LEMMA. *Let* (X_t) *be a one-dimensional Brownian motion. Then*

(a) $P(\overline{\lim}\ X_t = \infty) = 1$; *consequently* $P(\overline{\lim}\ X_t < \infty) = 0$.

(b) $P(\underline{\lim}\ X_t = -\infty) = 1$; *consequently* $P(\underline{\lim}\ X_t > -\infty) = 0$.

(c) $P(\underline{\lim}\ X_t = -\infty\ ;\ \overline{\lim}\ X_t = \infty) = 1$.

SIGNIFICANCE. By (c) almost every Brownian path assumes each value infinitely often.

Proof. $$\{\overline{\lim}\ X_t = \infty\} = \bigcap_{n=1}^{\infty} (\overline{\lim}\ X_t > n)$$

$$= \bigcap_{n=1}^{\infty} (\overline{\lim}_{\theta \text{ rational}}\ X_\theta > n) \text{ (by continuity of Brownian paths)}$$

First, note that

$$P_o[\sup_{0 \le s \le t} X(s) \le n] = 1 - P_o[\sup_{0 \le s \le t} X(s) > n]$$

$$= 1 - 2\, 1/\surd(2\pi t) \int_n^\infty e^{-y^2/2t}\, dy$$

$$= \surd(2/\pi t) \int_0^n e^{-y^2/2t}\, dy.$$

Therefore, for any x_o and t,

$$P[\sup_{t_o \le s \le t} X(s) \ge n\ |X(t_o) = x_o] = P_o[\sup_{o \le s \le t - t_o} X(s) \ge n - x_o]$$

(independent increments) which tends to 1 as $t \to \infty$. Consequently,

$$P_o[\sup_{t \geq t_o} X(t) \geq n] = EP[\sup_{t \geq t_o} X(t) \geq n \,|X(t_o)]$$

$$= E\,1 = 1.$$

In other words,

$$P_o[\limsup_{t \to \infty} X(t) \geq n] = 1$$

for every n. Thus

$$P(\overline{\lim}\ X_t = \infty) = 1.$$

(b) is clear if one notes that $w \to -w$ leaves the probability invariant.

(c) $P(\overline{\lim}\ X_t = \infty, \underline{\lim}\ X_t = -\infty)$

$$= P(\overline{\lim}\ X_t = \infty) - P(\underline{\lim}\ X_t > -\infty, \overline{\lim}\ X_t = \infty).$$

$$\geq 1 - P(\underline{\lim}\ X_t > -\infty)$$

$$= 1.$$

COROLLARY. *Let* (X_t) *be a d-dimensional Brownian motion. Then* $P(\overline{\lim}\ |X_t| = \infty) = 1.$

REMARK. If $d \geq 3$ we shall see later that $P(\operatorname{Lt}_{t\to\infty} |X_t| = \infty) = 1.$ i.e. almost every Brownian path "wanders" off to ∞.

THEOREM. *Almost all Brownian paths are of unbounded variation in any interval.*

Proof. Let I be any interval $[a,b]$ with $a < b$. For $n = 1,2,\ldots$ define

$$V_n(wQ_n) = \sum_{i=1}^{n} |w(t_i) - w(t_{i-1})| \quad (t_i = a + (b-a)i/n,\ i = 0,1,2,\ldots n),$$

The variation corresponding to the partition Q_n dividing $[a,b]$ into n equal parts. Let

$$U_n(w,Q_n) = \sum_{i=1}^{n} |(w(t_i) - w(t_{i-1})|^2 .$$

If

$$A_n(w,Q_n) \quad \sup_{1 \leq i \leq n} |w(t_i) - w(t_{i-1})| ,$$

then

$$A_n(w,Q_n)\, V_n(w,Q_n) \geq U_n(w,Q_n).$$

By continuity $\operatorname*{Lt}_{n \to \infty} A_n(w,Q_n) = 0.$

CLAIM. $\operatorname*{Lt}_{n \to \infty} E[(U_n(w,Q_n) - (b - a))^2] = 0.$

Proof. $E[(U_n - (b - a))^2]$

$$= E \left\{ \sum_{j=1}^{n} [(X_{t_j} - X_{t_{j-1}})^2 - (b-a/n)] \right\}^2$$

$$= E \, [(\sum (Z_j^2 - b-a/n))^2], \; Z_j = X_{t_j} - X_{t_{j-1}},$$

$$= n \, E[(Z_1^2 - b-a/n)^2]$$

(because Z_j are independent and identically distributed).

$$= n[E(Z_1^4) - (b-a/n)^2] = 2(b -a/n)^2 \to 0.$$

Thus a subsequence $U_{n_i} \to b-a$ almost everywhere. Since $A_{n_i} \to 0$ it follows that $V_{n_i}(w,Q_n) \to \infty$ almost everywhere. This completes the proof.

NOTE. {w: w is of bounded variation on [a,b]} can be s hown to be measurable if one proves

Exercise. Let f be continuous on [a,b] and define $V_n(f,Q_n)$ as above. Show that f is of bounded variation on [a,b] iff

$$\sup_{n=1,2,\ldots} V_n(f,Q_n) < \infty.$$

THEOREM. *Let* t *be any fixed real number in* $[0,\infty)$, D_t = {w: w *is differentiable at* t}. *Then* $P(D_t) = 0$.

Proof. The measurability of D_t follows from the following observation: if f is continuous then f is differentiable at t if and only if

$$\underset{\substack{r\to 0\\ r \text{ rational}}}{\mathrm{Lt}} \frac{f(t+r)-f(t)}{r}.$$

exists. Now

$$D_t \quad \bigcup_{m=1}^{\infty} \quad w: \left|\frac{w(t+h)-w(t)}{h}\right| \leq M, \quad \text{for all} \quad h \neq 0, \text{ rational}\}$$

and

$$P\left\{w: \left|\frac{X_{t+h}-X_t}{h}\right| \leq M \;\forall\; h \in Q,\; h \neq 0\right\} \leq 2 \inf_h \int_0^{M\sqrt{h}} \frac{1}{\sqrt{(2\pi)}} e^{-|y|^2/2} dy = 0$$

REMARK. A stronger result holds:

$$P\Big(\bigcup_{t\geq 0} D_t\Big) = 0.$$

Hint: $$\bigcup_{0\leq t\leq 1} D_t \quad \bigcup_{=1} \bigcup_{m=1} \bigcup_{n=m} \bigcup_{i=1}^{n+2} \quad \{w: w(\tfrac{k}{n})-w(\tfrac{k-1}{n})| \leq \tfrac{71}{n}\}$$
$k=i+1,i+2,i+3$

and

$$P\Big(\bigcup_{i=1}^{n+2} \quad \{w: w(\tfrac{k}{n})-w(\tfrac{k-1}{n})| \leq \tfrac{71}{n}\} \quad \text{const}/\sqrt{n}$$
$k=i+1,\ldots,i+3$

This construction is due to A. Dvoretski, P. Erdos & S. Kakutani.

10. DIRICHLET PROBLEM AND BROWNIAN MOTION

LET G BE ANY bounded open set in $\mathbb{R}^d$. Define the exit time $\tau_G(w)$ as follows:

$$\tau_G(w) = \{\inf \ t: w(t) \notin G\} .$$

If $w(0) \in G$, $\tau_G(w) = 0$; if $w(0) \in G$, $\tau_G(w)$ is the first time w escapes G or, equivalently, it is the first time that w hits the boundary ∂G of G. Clearly $\tau_G(w)$ is a stopping time. By definition $X_{\tau_G}(w) \in \partial G$, $\forall\, w$ and X_{τ_G} is a random variable. We can define a Borel probability measure on ∂G by

$$\begin{aligned} \Pi_G(x,\Gamma) &= P_x(X_{\tau_G} \in \Gamma) \\ &= \text{probability that } w \text{ hits } \Gamma. \end{aligned}$$

If f is a bounded, real-valued measurable function defined on ∂G, we define

$$u(x) = E_x(f(X_{\tau_G})) = \int_{\partial G} f(y)\, \Pi_G(x,dy)$$

where

$$E_x = E^{P_x'} .$$

In case G is a sphere centred around x, the exact form of $\Pi_G(x,\Gamma)$ is computable.

<u>THEOREM</u>. *Let* $S = S(0;r) = \{y \in \mathbb{R}^d : |y| < r\}$. *Then*

$$\Pi_S(0,\Gamma) = \frac{\text{surface area of } \Gamma}{\text{surface area of } S} .$$

Proof. The distributions $\{F_{t_1,\ldots,t_k}\}$ defining Brownian motion are invariant under rotations. Thus $\Pi_S(0,.)$ is a rotationally invariant probability measure. The result follows from the fact that the only probability measure (on the surface of a sphere) that is invariant under rotations is the normalised surface area.

THEOREM. *Let G be any bounded region, f a bounded measurable real valued function defined on ∂G. Define $u(x) = E_x(f(X_{\tau_G}))$. Then*

(i) u is measurable and bounded;

(ii) u has the mean value property; consequently,

(iii) u is harmonic in G.

Proof. (i) To prove this, it is enough to show that the mapping $x \to P_x(A)$ is measurable for every Borel set A.

Let $\mathcal{C} = \{A \in \mathcal{B} : x \to P_x(A)$ is measurable$\}$

It is clear that $\Pi^{-1}_{t_1,\ldots,t_k}(B) \in \mathcal{C}$, $\forall$ Borel set B in $\mathbb{R}^d \times .. \times \mathbb{R}^d$.

As $\mathcal{C}$ is a monotone class $\mathcal{C} = \mathcal{B}$.

(ii) Let S be any sphere with centre at x, and $S \subset G$.

Let $\tau = \tau_S$ denote the exit time through S. Clearly $\tau \leq \tau_G$.

By the strong Markov property,

$$u(X_\tau) = E(f(X_{\tau_G}) | \mathcal{F}_\tau).$$

Now

$$u(x) = E_x(f(X_{\tau_G})) = E_x(E(f(X_{\tau_G}) | \mathcal{F}_\tau)$$

$$= E_x(u(X_\tau)) = \int_{\partial S} u(y)\, \Pi_S(x,dy)$$

$$= \frac{1}{|S|} \int_{\partial S} u(y) dS \ ; \ |S| = \text{surface area of } S.$$

(iii) is a consequence of (i) and (ii). (See exercise below).

ercise.' Let u be a bounded measurable function in a region G tisfying the mean value property, i.e.

$$u(x) = \frac{1}{|S|} \int_{\partial S} u(y)\, dS$$

r every sphere S G. Then

(i) $u(x) = \frac{1}{\text{vol} S} \int_S u(y)\, dy.$

(ii) Using (i) show that u is continuous.

(iii) Using (i) and (ii) s how that u is harmonic.

We shall now solve the boundary value problem under suitable nditions on the region G.

EOREM. *Let* G, f, u *be as in the previous theorem. Further suppose at*

(i) f *is continuous ;*

(ii) G *satisfies the exterior cone condition at every point* ∂G, *i.e. for each* $y \in \partial G$ *there exists a cone* C_h *with vertex the point* y *of height* h *and such that* $C_h - \{y\} \subset$ *exterior of* G.

en

$$\lim_{x \to y, x \in G} u(x) = f(y), \forall y \in \partial G.$$

Proof. Step 1.

$$P_y \{w: w(0) = y, w \text{ remains in } \overline{G} \text{ for some positive time}\} = 0$$

Let $A_n = \{w: w(0) = y, w(s) \quad \overline{G} \text{ for } 0 \leq s \leq 1/n\}$,

$$B_n = \Omega - A_n, \; A = \bigcup_{n=1}^{\infty} A_n, \; B = \bigcap_{n=1}^{\infty} B_n.$$

As A_n's are increasing, B_n's are decreasing and $B_n \in \mathcal{F}_{1/n}$ so that $B \in \mathcal{F}_{o+}$. We s how that $P(B) > 0$, so that by Bluementhal's zero-one law, $P(B) = 1$, i.e. $P(A) = 0$.

$$P_y(B) = \lim_{n \to \infty} P_y(B_n) \geq \overline{\lim_{n \to \infty}} P_y \{w: w(0) = y, w(\tfrac{1}{2n}) \in C_h - \{y\}\}$$

Thus

$$P_y(b) \geq \overline{\lim} \int_{C_h - \{y\}} 1/\sqrt{(2\pi/2n)^d} \exp(-|z-y|^2 /2/2n)\, dz$$

$$= \int_{C_\infty} 1/\sqrt{(2\pi)}\, e^{-|y|^2/2}\, dy,$$

where C_∞ is the cone of infinite height obtained from C_h. Thus $P_y(B) > 0$.

Step 2. If C is closed then the mapping $x \to P_x(C)$ is upper semi-continuous.

For, denote by χ_C the indicator function of C. As C is closed (in a metric space) $\exists$ a sequence of continuous functions f_n decreasing to χ_C such that $0 \leq f_n \leq 1$. Thus $E_x(f_n)$ decreases to $E_x(\chi_C) = P_x(C)$. Clearly $x \to E_x(F_n)$ is continuous. The result follows from the fact that the infimum of any collection of continuous functions is upper semi-continuous.

Step 3. Let $\delta > 0$,

$$N(y;\delta) = \{z \in \partial G: |z-y| < \delta\},$$

$$B_\delta = \{w: w(0) \in G, X_{\tau_G}(w) \in \partial G - N(y;\delta)\},$$

i.e. B_δ consists of trajectories w hich start at a point of G and escape for the first time through ∂G at a point not in $N(y;\delta)$ If $C_\delta = \overline{B}_\delta$, then

$$C_\delta \cap \{w: w(0) = y\} \subset A \cap \{w: w(0) = y\}$$

where A is as in Step 1.

For, supp ose $w \in C_\delta \cap \{w: w(0) = y\}$. Then there exists $w_n \in B_\delta$ such that $w_n \to w$ uniformly on compact sets. If $w \notin A \cap \{w: w(0) = y\}$ there exists $\varepsilon > 0$ such that $w(t) \in \overline{G}\ \forall\ t$ in $(0,\varepsilon]$. Let $\delta^* = \inf_{0 \le t \le \varepsilon} d(w(t), G - N(y,\delta))$. Then $\delta^* > 0$.

If $t_n = \tau_G(w_n)$ and t_n does not converge to 0, then there exists a subsequence, again denoted b y t_n, such that $t_n \ge k\varepsilon > 0$ for some $k \in (0,1)$. Since $w_n(k\varepsilon) \in \overline{G}$ and $w_n(k\varepsilon), w(k\varepsilon) \in \overline{G}$, a contradiction. Thus we can assume that t_n converges to 0 and also that $\varepsilon \ge t_n\ \forall\ n$, But then

$$(*) \qquad |w_n(t_n) - w(t_n)| \ge \delta^*.$$

However, as w_n converges to w uniformly on $[0,\varepsilon]$,

$$w_n(t_n) - w(t_n) \to w(0) - w(0) = 0$$

contradicting (*). Thus $w \in A \quad \{w: w(0) = y\}$.

Step 4. $\lim_{x \to y, x \in G} P_x(B_\delta) = 0.$

For, $\overline{\lim}_{x \to y} P_x(B_\delta) \leq \overline{\lim}_{x \to y} P_x(C_\delta) \leq P_y(C_\delta)$ (by Step 2)

$$= P_y(C_\delta \cap \{w: w(0) = y\})$$

$$\leq P_y(A) \quad \text{(by Step 3)}$$

$$= 0.$$

Step 5. $|u(x) - f(y)| = |\int_\Omega f(X_{\tau_G}(w))dP_x(w) - \int_\Omega f(y)dP_x(w)|$

$$\leq \int_{\Omega - B_\delta} |f(X_{\tau_G}(w)) - f(y)|dP_x(w) + |\int_{B_\delta} (f(X_{\tau_G}(w)) - f(y))dP_x($$

$$\leq \int_{\Omega - B_\delta} |f(X_{\tau_G}(w)) - f(y)|dP_x(w) + 2\|f\|_\infty P_x(B_\delta)$$

and the right hand side converges to 0 as $x \to y$ (byy Step 4 and t fact that f is continuous). This proves the theorem.

REMARK. The theorem is local.

THEOREM. *Let* $G = \{y \in \mathbb{R}^d: \delta < |y| < R\}$, f *any continuous functio on* $\partial G = \{|y| = \delta\} \cup \{|y| = R\}$. *If* u *is any harmonic function in* G *with boundary values* f, *then* $u(x) = E_x(f(X_{\tau_G}))$.

Proof. Clearly G has the exterior cone property. Thus, if $v(x)$ $E_x(f(X_{\tau_G}))$, then v is harmonic in G and has boundary values f (by the previous theorem). The result follows from the uniqueness of the solution of the Dirichlet problem for the Laplacian operator.

The function $f = 0$ on $|y| = R$ and $f = 1$ on $|y| = \delta$ is of special interest. Denote by $U_{\delta,1}^{R,0}$ the corresponding solution of the Dirichlet problem.

Exercise (i). If $d = 2$ then

$$U_{\delta,1}^{R,0}(x) = \frac{\log R - \log |x|}{\log R - \log \delta}, \forall x \in G.$$

(ii). If $d \geq 3$ then

$$U_{\delta,1}^{R,0}(x) = \frac{|x|^{-n+2} - R^{-n+2}}{\delta^{-n+2} - R^{-n+2}}.$$

Case (i): $d = 2$. Then

$$\frac{\log R - \log |x|}{\log R - \log \delta} = U_{\delta,1}^{R,0}(x).$$

Now,

$$E_x(f(X_{\tau_G})) = \int_{|y|=\delta} \Pi_G(x,dy) = P_x(|X_{\tau_G}| = \delta),$$

i.e.

$$\frac{\log R - \log |x|}{\log R - \log \delta} = P_x(|X_{\tau_G}| = \delta)$$

$$= P_x \text{ (the particle hits } |y| = \delta \text{ before it hits } |y| = R).$$

Fix R and let $\delta \to 0$; then

$$0 = P_x \text{ (the particle hits } 0 \text{ before hitting } |y| = R).$$

Let R take values $1,2,3,\ldots$, then $0 = P_x$ (the particle hits 0 before hitting any of the circles $|y| = N$). Recalling that

$$P_x(\overline{\lim}\, |X_t| = \infty) = 1,$$

we get

PROPOSITION. *A two-dimensional Brownian motion does not visit a point.*

Next, keep δ fixed and let $R \to \infty$, then,

$$1 = P_x(|w(t)| = \delta \text{ for some time } t > 0).$$

Since any time t can be taken as the starting time for the Brownian motion, we have

PROPOSITION. *Two-dimensional Brownian motion has the recurrence property.*

Case (ii): $d \geq 3$. In this case

$$P_x(w : w \text{ hits } |y| = \delta \text{ before it hits } |y| = R)$$

$$= (1/|x|^{n-2} - 1/R^{n-2}) \,/\, (1/\delta^{n-2} - 1/R^{n-2}).$$

Letting $R \to \infty$ we get

$$P_x(w: w \text{ hits } |y| = \delta) = (\delta/|x|)^{n-2}$$

which lies strictly between 0 and 1. Fixing δ and letting $|x| \to \infty$, we have

PROPOSITION. *If the particle start at a point far away from 0 then it has very little chance of hitting the circle $|y| = \delta$.*

If $|x| \leq \delta$, then

$$P(w \text{ hits } S_\delta) = 1 \quad \text{where } S_\delta = \{y \in R^d : |y| = \delta\}.$$

Let

$$V_\delta(x) = (\delta/|x|)^{n-2} \quad \text{for } |x| \geq \delta.$$

In view of the above result it is natural to extend V_δ to all space by putting $V_\delta(x) = 1$ for $|x| \leq \delta$. As Brownian motion has the Markov property

$$P_x\{w: w \text{ hits } S_\delta \text{ after time } t\}$$

$$= \int V_\delta(y)\, 1/\sqrt{(2\pi t)}^d \exp -|y|^2/2t \, dy \to 0 \quad \text{as } t \to +\infty.$$

Thus $P(w$ hits S_δ for arbitrarily large $t) = 0$.

In other words, $P\left(w: \underline{\lim}_{t \to \infty} |w(t)| \geq \delta\right) = 1$. As this is true $\forall\, \delta > 0$, we get the following important result.

<u>PROPOSITION</u>. $P\left(\underline{\lim}_{t \to \infty} |w(t)| = \infty\right) = 1$,

i.e. *for* $d \geq 3$, *the Brownian particles wander away to* $+\infty$.

11. STOCHASTIC INTEGRATION

LET $\{X_t : t \geq 0\}$ BE A one-dimensional Brownian motion. We want first to define integrals of the type $\int_0^\infty f(s)\ dX(s)$ for real functions $f \in L^2[0,\infty)$. If $X(s,w)$ is of bounded variation almost everywhere then we can give a meaning to $\int_0^\infty f(s)\ dX(s,w) = g(w)$. However, since $X(s,w)$ is not bounded variation almost everywhere, $g(w)$ is not defined in the usual sense.

In order to define $g(w) = \int_0^\infty f(s)\ dX(s,w)$ proceed as follows.

Let f be a step function of the following type:

$$f = \sum_{i=1}^{n} a_i \chi_{[t_i,t_{i+1})},\ 0 \leq t_1 < t_2 < .. < t_{n+1}.$$

We naturally define

$$g(w) = \int_0^\infty f(s)\ d\,X(s,w) = \sum_{i=1}^{n} a_i (X_{t_{i+1}}(w) - X_{t_i}(w))$$

$$= \sum_{i=1}^{n} a_i (w(t_{i+1}) - w(t_i)).$$

g satisfies the following properties:

(i) g is a random variable;

(ii) $E(g) = 0$; $E(g^2) = \sum a_i^2 (t_{i+1} - t_i) = \|f\|_2$.

This follows from the facts that (a) $X_{t_{i+1}} - X_{t_i}$ is a normal random variable with mean 0 and variance $(t_{i+1} - t_i)$ and (b) $X_{t_{i+1}} - X_{t_i}$ are independent increments, i.e. we have

$$E\left(\int_0^\infty f\, dX\right) = 0, \quad E\left(\left|\int_0^\infty f dX\right|^2\right) = \|f\|_2^2.$$

Exercise 1. If

$$f = \sum_{i=1}^{n} a_i \chi_{[t_i, t_{i+1})}, \quad 0 \leq t_1 < \ldots < t_{n+1},$$

$$g = \sum_{i=1}^{m} b_i \chi_{[s_i, s_{i+1})}, \quad 0 \leq s_1 < \ldots < s_{m+1},$$

Show that

$$\int_0^\infty (f+g)\, dX(s,w) = \int_0^\infty f dX(s,w) + \int_0^\infty g dX(s,w)$$

and

$$\int_0^\infty (\alpha f) dX(s,w) = \alpha \int_0^\infty f dX(s,w), \ \forall\, \alpha \in \mathbb{R}.$$

REMARK. The mapping $f \to \int_0^\infty f dX$ is therefore a linear $L^2_{\mathbb{R}}$ -isometry of the space S of all simple functions of the type

$$\sum_{i=1}^{n} a_i \chi_{[t_i, t_{i+1})}, \ (0 \leq t_1 < \ldots < t_{n+1})$$

into $L^2(\Omega, \mathcal{B}, P)$.

Exercise 2. Show that S is a dense subspace of $L^2[0,\infty)$.

Hint: $C_c[0,\infty)$, i.e. the set of all continuous functions with compact support, is dense in $L^2[0,\infty)$. Show that S contains the closure of $C_c[0,\infty)$.

REMARK. The mapping $f \to \int_0^\infty f dX$ can now be uni quely extended as an isometry of $L^2[0,\infty)$ into $L^2(\Omega, \mathcal{B}, P)$.

Next we define integrals of the type

$$g(w) = \int_0^t X(s,w)\, dX\,(s,w)$$

Put $t = 1$ (the general case can be dealt with similarly). It seems natural to define

$$(*) \qquad \int_0^1 X(s,w)dX(s) = \underset{\sup |t_j - t_{j-1}| \to 0}{\mathrm{Lt}} \sum_{j=1}^{n} X(\xi j)(X(t_j) - X(t_{j-1})$$

where $0 = t_o < t_1 < \ldots < t_n = 1$ is a partition of $[0,1]$ with $t_{j-1} \leq \xi_j \leq t_j$. In general the limit on the right hand side may not exist. Even if it exists it may happen th at depending on the choice of ξ_j, we may obtain different limits. To consider an example we choose $\xi_j = t_j$ and then $\xi_j = t_{j-1}$ and compute the right hand side of (*). If $\xi_j = t_{j-1}$,

$$\sum_{j=1}^{n} X_{\xi_j}(X_{t_j} - X_{t_{j-1}}) = \sum_{j=1}^{n} X_{t_{j-1}} -(X_{t_j} - X_{t_{j-1}})$$

$$= \frac{1}{2} \sum_{j=1}^{n} (X_{t_j}) - (X_{t_{j-1}})) - \frac{1}{2} \sum_{j=1}^{n} (X_{t_j} - X_{t_{j-1}})$$

$$\frac{1}{2} [X^2(1) - X^2(0)] - \frac{1}{2} \quad \text{as } n \to \infty, \text{ and } \sup |t_j - t_{j-1}| \to 0,$$

arguing as in the proof of the result that Brownian motion is not of bounded variation. If $\xi_j = t_j$,

$$\underset{\substack{n \to \infty \\ \mathrm{Sup}\, |t_j - t_{j-1}| \to 0}}{\mathrm{Lt}} \sum_{j=1}^{n} X_{t_j} (X_{t_j} - X_{t_{j-1}}) = 1/2\ X\ (1) - 1/2\ X\ (0) + 1/2.$$

Thus we get different answers depending on the choice of ξ_j and hence one has to be very careful in defining the integral. It turns out that the choice of $\xi_j = t_{j-1}$ is more appropriate in the definition of the integral and gives better results.

REMARK. The limit in (*) should be understood in the sense of convergence probability.

Exercise 3. Let $0 \le a < b$. Show that the "left integral" $(\xi_j = t_{j-1})$ is given by

$$L \int_a^b X(s)\, dX(s) = \frac{X^2(b) - X^2(a) - (b-a)}{2}$$

and the "right integral" $(\xi_j = t_j)$ is given by

$$R \int_a^b X(s)\, dX(s) = \frac{X^2(b) - X^2(a) + (b-a)}{2} .$$

We now take up the general theory of s tochastic integration. To motivate the definitions which follow let us consider a d-dimensional Brownian motion $\{\beta(t): t \ge 0\}$. We have

$$E[\beta(t+s) - \beta(t) \in A \mid \mathcal{F}_t] = \int_A 1/\sqrt{(2\pi s)}\, e^{-|y|^2/2s}\, dy .$$

Thus

$$E[f(\beta(t+s) - \beta(t)) \mid \mathcal{F}_t] = \int f(y)\, 1/\sqrt{(2\pi s)}\, e^{-|y|^2/2s}\, dy.$$

In particular, if $f(x) = e^{ix.u}$,

$$E[e^{iu}\, (\beta(t+s) - \beta(t)) \mid \mathcal{F}_t] = \int e^{iu.y}\, 1/\sqrt{(2\pi s)}\, e^{-|y|^2/2s}\, dy$$

$$= e^{-s|u|^2/2}.$$

Thus

$$E[e^{iu.\beta(t+s)}|\mathcal{F}_t] = e^{iu.\beta(t)}\ e^{-s|u|^2/2},$$

or,

$$E[e^{iu.\beta(t+s)+(t+s)|u|^2/2}|\mathcal{F}_t] = e^{iu.\beta(t)+t|u|^2/2}.$$

Replacing iu by θ we get

$$E[e^{\theta.\beta(s\ )-|s\theta|^2/2}|\mathcal{F}_t] = e^{\theta.\beta(t)-t|\theta|^2/2},\quad s > t, \forall\ \theta.$$

It is clear that $e^{\theta.\beta(t)\ -\ t|\theta|^2/2}$ is $\mathcal{F}_t$ -measurable and a simple calculation gives

$$E(\ e^{\theta.\beta(t)\ -\ |\theta|^2 t/2|}) < \infty\quad \forall\ \theta.$$

We thus have

<u>THEOREM</u>. *If* $\{\beta(t): t \geq 0\}$ *is a d-dimensional Brownian motion then* $\exp\ [\theta.\beta(t) - |\theta|^2 t/2]$ *is a Martingale relative to* $\mathcal{F}_t$, *the* σ *-algebra generated by* $(\beta(s): s \leq t)$.

<u>DEFINITION</u>. Let $(\Omega, \mathcal{B}, P)$ be a probability space $(\mathcal{F}_t)_{t \geq 0}$ an increasing family of sub- σ -algebras of $\mathcal{F}$ with $\mathcal{F} = \sigma(\bigcup_{t \geq 0} \mathcal{F}_t)$. Let

(i) $a: [0,\infty)\ x\Omega \to [0,\infty)$ be bounded and progressively measurable;

(ii) $b: [0,\infty)\ x\Omega \to \mathbb{R}$ be bounded and progressively measurable;

(iii) $X: [0,\infty)\ x\Omega \to \mathbb{R}$ be progressively measurable, right continuous on $[0,\infty), \forall\ w \in \Omega$, and continuous on $[0,\infty)$ almost everywhere on Ω;

(iv) $Z_t(w) = e^{\theta X(t,w) - \theta \int_0^t b(s,w)ds - \frac{\theta^2}{2}\int_0^t a(s,w)ds}$

be a Martingale relative to $(\mathcal{F}_t)_{t \geq 0}$.

Then $X(t,w)$ is called an Ito process corresponding to the parameters b and a and we write $X_t \in I\,[b,a]$.

N.B. The progressive measurability of $X \Rightarrow X_t$ is $\mathcal{F}_t$-measurable.

Example. If $\{\beta(t): t \geq 0\}$ is a Brownian motion, then $X(t,w)=\beta_t(w)$ is an Ito process corresponding to parameters 0 and 1. (i) and (ii) are obvious. (iii) follows by right continuity of β_t and measurability of β_t relative to $\mathcal{F}_t$ and (iv) is proved in the previous theorem.

Exercise 4. Show that $Z_t(w)$ defined in (iv) is $\mathcal{F}_t$ -measurable and progressively measurable.

[Hint: (i) Z_t is right continuous.

(ii) Use Fubini's theorem to prove measurability].

REMARK. If we put $Y(t,w) = X(t,w) - \int_0^t b(s,w)ds$ then $Y(t,w)$ is progressively measurable and $Y(t,w)$ is an Ito process corresponding to the parameters $0,a$. Thus we need only consider integrals of the type $\int_0^t f(s,w)\,dY(s,w)$ and *define*

$$\int_0^t f(s,w)dX(s,w) = \int_0^t f(s,w)dY(s,w) + \int_0^t f(s,w)b(s,w)ds.$$

(Note that *formally* we have $dY = dX - dbt$).

LEMMA. *If* $Y(t,w) \in I[0,a]$, *then*

$$Y(t,w) \text{ and } Y^2(t,w) - \int_0^t a(s,w)ds$$

are Martingales relative to $(\mathcal{F}_t)$.

Proof. To motivate the arguments which follow, we first give a formal proof. Let

$$Y_\theta(t) = e^{\theta Y(t,w) - \frac{\theta^2}{2}\int_0^t a(s,w)\,ds}.$$

Then $Y_\theta(t)$ is a martingale, $\forall\ \theta$. Therefore $\frac{Y_\theta - 1}{\theta}$ is a Martingale, $\forall\ \theta$. Hence (formally),

$$\lim_{\theta \to 0} \frac{Y_\theta - 1}{\theta} = Y'_\theta\big|_{\theta=0}$$

is a Martingale.

Step 1. $Y(t,.) \in L^k(\Omega,\mathcal{F},P)$, $k = 0,1,2,\ldots$ and $\forall\ t$.

In fact, for every real θ, $Y_\theta(t)$ is a Martingale and hence $E(Y_\theta) < \infty$. Since a is bounded this means that

$$E(e^{\theta Y(t,.)}) < \infty,\ \forall\ \theta.$$

Taking $\theta = 1$ and -1 we conclude that $E(e^{|Y|}) < \infty$ and hence $E(|Y|^k) < \infty$, $\forall\ k = 0,1,2,\ldots$. Since Y is an Ito process we also get

$$\sup_{|\theta| \le \alpha} E\left(\left|e^{Y(t,.) - \frac{\theta^2}{2}\int_0^t a\,ds}\right|^k\right) < \infty$$

$\forall\ k$ and for every $\alpha > 0$.

<u>Step 2</u>. Let $X_\theta(t) = [Y(t,.) - \theta\int_0^t a ds]Y(t) = \frac{d}{d\theta} Y_\theta(t,.)$.

Define

$$\phi_A(\theta) = \int_A (X_\theta(t,.) - X_\theta(s,.))\, dP(w)$$

where $t > s$, $A \in \mathcal{F}_s$. Then

$$\int_{\theta_1}^{\theta_2} \phi_A(\theta)\, d\theta = \int_{\theta_1}^{\theta_2} \int_A [X_\theta(t,.) - X_\theta(s,.)]\, dP(w)\, d\theta.$$

Since a is bounded, $\sup_{|\theta| \leq \alpha} E([Y_\theta(t,.)]^k) < \infty$, and $E(|Y|^k) < \infty, \forall\, k$; we can use Fubini's theorem to get

$$\int_{\theta_1}^{\theta_2} \phi_A(\theta)\, d\theta = \int_A \int_{\theta_1}^{\theta_2} [X_\theta(t,.) - X_\theta(s,.)]\, d\theta\, dP(w).$$

or

$$\int_{\theta_1}^{\theta_2} \phi_A(\theta)\, d\theta = \int_A Y_{\theta_2}(t,.) - Y_{\theta_1}(t,.) dP(w) - \int_A Y_{\theta_1}(s,.) - Y_{\theta_1}(s,.) dP(w).$$

Let $A \in \mathcal{F}_s$ and $t > s$; then, since Y is a Martingale,

$$\int_{\theta_1}^{\theta_2} \phi_A(\theta)\, d\theta = 0.$$

This is true $\forall\, \theta_1 < \theta_2$ and since $\phi_A(\theta)$ is a continuous function of θ, we conclude that

$$\phi_A(\theta) = 0, \quad \forall\, \theta.$$

In particular, $\phi_A(\theta) = 0$ which means that

$$\int_A Y(t,.) dP(w) = \int_A Y(s,.) dP(w), \ \forall\, A \in \mathcal{F}_s, \ t > s,$$

i.e., $Y(t)$ is a Martingale relative to $(\Omega, \mathcal{F}_t, P)$.

To prove the second part we put

$$Z_\theta(t,.) = \frac{d^2}{d\theta^2} Y_\theta(t)$$

and

$$\psi_A(\theta) = \int_A \{Z_\theta(t,.) - Z_\theta(s,.)\} \, dP(w).$$

Then, by Fubini,

$$\int_{\theta_1}^{\theta_2} \psi_A(\theta) \, d\theta = \int_A \int_{\theta_1}^{\theta_2} Z_\theta(t,.)-Z_\theta(s,.) d\theta \, dP(w).$$

or,

$$\int_{\theta_1}^{\theta_2} \psi_A(\theta) \, d\theta = \phi_A(\theta_2) - \phi_A(\theta_1)$$

$$= 0 \text{ if } A \in \mathcal{F}_s, \; t > s.$$

Therefore

$$\psi_A(\theta) = 0, \quad \forall \theta.$$

In particular, $\psi_A(\theta) = 0$ implies that

$$Y^2(t,w) - \int_0^t a(s,w) \, ds$$

is an $(\Omega, \mathcal{F}_t, P)$ Martingale. This completes the proof of lemma 1.

<u>DEFINITION</u>. A function $\theta: [0,\infty) \times \Omega \to \mathbb{R}$ is called simple if there exist reals $s_o, s_1, \ldots, s_n, \ldots$

$$0 \leq s_o < s_1 < \ldots < s_n \ldots < \infty,$$

s_n increasing to $+\infty$ and

$$\theta(s,w) = \theta_j(w)$$

if $s \in [s_j, s_{j+1})$, where $\theta_j(w)$ is $\mathcal{F}_{s_j}$-measurable and bounded.

<u>DEFINITION</u>. Let $\theta\colon [0,\infty) \times \Omega \to \mathbb{R}$ be a simple function and $Y(t,w) \in I[0,a]$. We define the stochastic integral of θ with respect to Y, denoted

$$\int_0^t \theta(s,w)\, dY\,(s,w)),$$

by

$$\xi(t,w) = \int_0^t \theta(s,w)dY\,(s,w) = \sum_{j=1}^{k} \theta_{j-1}(w)[Y(s_j,w) - Y(s_{j-1},w)] + \theta_k(w)[Y(t,w) - Y(s_k,w)].$$

<u>LEMMA 2</u>. *Let* $\sigma\colon [0,\infty) \times \Omega \to \mathbb{R}$ *be a simple function and* $Y(t,w) \in I[0,a]$. *Then*

$$\xi(t,w) = \int_0^t \sigma(s,w)dY(s,w) \in I[0,a\sigma^2].$$

<u>Proof.</u> (i) By definition, σ is right continuous and $\sigma(t,w)$ is $\mathcal{F}_t$-measurable; hence it is progressively measurable. Since a is progressively measurable and bounded

$$a\sigma^2\colon [0,\infty) \times \Omega \to [0,\infty)$$

is progressively measurable and bounded.

(ii) From the definition of ξ it is clear that $\xi(t,.)$ is right continuous, continuous almost everywhere and $\mathcal{F}_t$- measurable therefore ξ is progressively measurable.

(iii) $$Z_t(w) = e^{[\theta\xi(t,w) - \frac{\theta^2}{2}\int_0^t a\sigma^2 \, ds]}$$

is clearly $\mathcal{F}_t$ - measurable $\forall\ \theta$. We show that

$$E(Z_t) < \infty,\ \forall\ t \text{ and } E(Z_{t_2}|\mathcal{F}_{t_1})=Z_{t_1} \text{ if } t_1 < t_2.$$

We can assume without loss of generality that $\theta = 1$ (if $\theta \neq 1$ we replace σ by $\theta\sigma$). Therefore

$$Z_t(w) = e^{[\xi(t,w) - \int_0^t a\sigma^2 \, ds]}.$$

Since a and σ are bounded uniformly on $[0,t]$, it is enough to show that $E(e^{\xi(t,w)}) < \infty$. B y definition,

$$\xi(t,w) = \sum_{j=1}^{k} \theta_{j-1}(w)[Y(s_j,w)-Y(s_{j-1},w)]+ \theta_k(w)(Y\ (t,w)- Y(s_k,w)).$$

The result $E(e^{\xi(t,w)}) < \infty$ will follow from the generalised Holder's inequality provided we show that

$$E(e^{\theta(w)[Y(t,w)-Y(\ s,w)]}) < \infty$$

for every bounded function θ which is $\mathcal{F}_s$ -measurable. Now

$$E(e^{\theta[Y(t,.)-Y(s,.)]}|\mathcal{F}_s)= \text{constant for every constant } \theta,$$

since $Y \in I[0,a]$. Therefore

$$E(e^{\theta(w)[Y(t,.)-Y(\ s,.)]}|\ \mathcal{F}_s) = \text{constant}$$

for every θ which is bounded and $\mathcal{F}_s$-measurable. Thus

$$E(e^{\theta(w)[Y(t,.)-Y(s,.)]}) < \infty.$$

This proves that $E(Z_t(w)) < \infty$.

Finally we show that

$$E(Z_{t_2} \mid \mathcal{F}_{t_1}) = Z_{t_1}(w), \quad \text{if} \quad t_1 < t_2.$$

Consider first the case when t_1 and t_2 are in the same interval $[s_k, s_{k+1})$. Then

$$\xi(t_2,w)=\xi(t_1,w)+\sigma_k(w)[Y(t_2,w)-Y(t_1,w)] \text{ (see definition)},$$

$$\int_0^{t_2} a\sigma^2(s,w)ds = \int_0^{t_1} a\sigma^2(s,w)ds + \int_{t_1}^{t_2} a\sigma^2(s,w)ds.$$

Therefore

$$E(Z_{t_2}(w) \mid \mathcal{F}_{t_1})=Z_{t_1}(w)E(\exp[\theta\sigma_k(w)[Y(t_2,w)-Y(t_1,w)]- \frac{\theta^2}{2} \int_{t_1}^{t_2} a\sigma^2 ds) \mid \mathcal{F}_{t_1})$$

as $Y \in I[0,a]$.

$$(*) \qquad E(\exp[\theta(Y(t_2,w)-T(t_1,w))- \frac{\theta^2}{2} \int_{t_1}^{t_2} a(s\ ,w)ds] \mid \mathcal{F}_{t_1}) = 1$$

and since $\sigma_k(w)$ is $\mathcal{F}_{t_1}$-measurable (*) remains valid if θ is replaced by $\theta\sigma_k$ Thus

$$E(Z_{t_2} \mid \mathcal{F}_{t_1}) = Z_{t_1}(w) .$$

The general case follows if we use the identity

$$E(E(X \mid \mathcal{C}_1) \mid \mathcal{C}_2) = E(X \mid \mathcal{C}_2) \quad \text{for} \quad \mathcal{C}_2 \subset \mathcal{C}_1.$$

Thus Z_t is a Martingale and $\xi(t,w) \in I[0,a\sigma^2]$.

COROLLARY. (i) $\xi(t,w)$ *is a martingale;* $E(\xi(t,w)) = 0$;

(ii) $\xi^2(t,w) - \int_0^t a\sigma^2\, ds$

is a Martingale with

$$E(\xi^2(t,w)) = E(\int_0^t a\sigma^2(s,w)\, ds.$$

Proof. Follows from Lemma 1.

LEMMA 3. *Let* $\sigma(s,w)$ *be progressively measurable such that for each* t*,*

$$E(\int_0^t \sigma^2(s,w)\, ds) < \infty.$$

Then there exists a sequence $\sigma_n(s,w)$ *of simple functions such that*

$$\lim_{n\to\infty} E(\int_0^t |\sigma_n(s,w) - \sigma(s,w)|^2\, ds) = 0\ .$$

Proof. We may assume that σ is bounded, for if $\sigma_N = \sigma$ for $|\sigma| \leq N$ and 0 if $|\sigma| > N$, then $\sigma_n \to \sigma$, $\forall\ (s,w) \in [0,t] \times \Omega$. σ_N is progressively measurable and $|\sigma_n - \sigma|^2 \leq 4|\sigma|^2$. By hypothesis $\sigma \in L\ ([0,t]\colon \Omega)$.

Therefore $E(\int_0^t |\sigma_n - \sigma| ds) \to 0$, by dominated convergence. Further, we can also assume that σ is continuous. For, if σ is bounded, define

$$\sigma_h(t,w) = 1/h \int_{(t-h)\vee 0}^{t} \sigma(s,w) ds.$$

σ_h is continuous in t and $\mathcal{F}_t$-measurable and hence progressively measurable. Also by Lebesgue's theorem

$$\sigma_h(t,w) \to \sigma(t,w),\ \text{as}\ h \to 0,\ \forall\ t,w.$$

Since σ is bounded by C, σ_h is also bounded by C. Thus

$$E(\int_0^t |\sigma_h(s,w) - \sigma(s,w)|^2 \, ds) \to 0.$$

(by dominated convergence). If σ is continuous, bounded and progressively measurable, then

$$\sigma_n(s,w) = \sigma(\frac{[ns]}{n}, w)$$

is progressively measurable, bounded and simple. But

$$\underset{n \to \infty}{\text{Lt}} \; \sigma_n(s,w) = \sigma(s \ ,w).$$

Thus by dominated convergence

$$E(\int_0^t |\sigma_n - \sigma|^2 \, ds) \to 0 \quad \text{as} \quad n \to \infty.$$

THEOREM. *Let* $\sigma(s,w)$ *be progressively measurable, such that* $E(\int_0^t \sigma^2(s,w) \, ds) < \infty$ *for each* $t > 0$. *Let* (σ_n) *be simple approximations to* σ *as in Lemma 3. Put*

$$\xi_n(t,w) = \int_0^t \sigma_n(s,w) \, dY(s,w)$$

where $Y \in I[0,a]$. *Then*

(i) $\underset{n \to \infty}{\text{Lt}} \; \xi_n(t,w)$ *exists uniformly in probability, i.e. there exists an almost surely continuous* $\xi(t,w)$ *such that*

$$\underset{n \to \infty}{\text{Lt}} \; P(\sup_{0 \le t \le T} |\xi_n(t,w) - \xi(t,w)| \ge \varepsilon) = 0$$

for each $\varepsilon > 0$ *and for each* T. *Moreover,* ξ *is independent of the sequence* (σ_o).

(ii) *The map* $\sigma \to \xi$ *is linear.*

(iii) $\xi(t,w)$ *and* $\xi^2(t,w) - \int_0^t a\sigma^2\, ds$ *are Martingales.*

(iv) *If* σ *is bounded,* $\xi \in I[0,a\sigma^2]$.

Proof. (i) It is easily seen that for simple functions the stochastic integral is linear. Therefore

$$(\xi_n - \xi_m)(t,w) = \int_0^t (\sigma_n - \sigma_m)(s,w)dY(s,w).$$

Since $\xi_n - \xi_m$ is an almost surely continuous martingale

$$P\left(\sup_{0 \le t \le T} |\xi_n(t,w) - \xi_m(t,w)| \ge \varepsilon\right) \le \frac{1}{\varepsilon^2} E[(\xi_n - \xi_m)^2(T,w)]$$

This is a consequence of Kolmogorov inequality (See Appendix). Since

$$(\xi_n - \xi_m)^2 - \int_0^t a(\sigma_n - \sigma_m)^2\, ds$$

is a Martingale, and a is bounded,

$$(*) \qquad E[(\xi_n - \xi_m)^2(T,w)] = E\left(\int_0^T (\sigma_n - \sigma_m)^2 a\, ds\right).$$

$$\le \text{const}\ \frac{1}{\varepsilon^2} E\left(\int_0^T (\sigma_n - \sigma_m)^2\, ds\right).$$

Therefore

$$\operatorname*{Lt}_{n,m \to \infty} E[(\xi_n - \xi_m)^2(T,w)] = 0.$$

Thus $(\xi_n - \xi_m)$ is uniformly Cauchy in probability. Therefore there exists a progressively measurable ξ such that

$$\operatorname*{Lt}_{n \to \infty} P\left(\sup_{0 \le t \le T} |\xi_n(t,w) - \xi(t,w)| \ge \varepsilon\right) = 0, \ \forall \varepsilon > 0, \ \forall T.$$

It can be shown that ξ is almost surely continuous .

If (σ_n) and (σ'_n) are two sequences of simple functions approximating σ, then (*) shows that

$$E[(\xi_n - \xi'_n)^2 (T,w)] \to 0.$$

Thus

$$\mathop{Lt}_n \xi_n = \mathop{Lt}_n \xi'_n ,$$

i.e. ξ is independent of (σ_n).

(ii) is obvious.

(iii) (*) shows that $\xi_n \to \xi$ in L and therefore $\xi_n(t,.) \to \xi(t,.)$ in L^1 for each fixed t. Since $\xi_n(t,w)$ is a martingale for each n, $\xi(t,w)$ is a martingale.

(iv) $\xi_n^2(t,w) - \int_0^t a\sigma_n^2 \, ds$ is a martingale for each n.

Since $\xi_n(t,w) \to \xi(t,w)$ in L^2 for each fixed t and

$$\xi_n^2(t,w) \to \xi^2(t,w) \text{ in } L^1 \text{ for each fixed } t.$$

For $\xi_n^2(t,w) - \xi^2(t,w) = (\xi_n - \xi)(\xi_n + \xi)$ and using Hölder's inequality, we get the result.

Similarly, since

$$\sigma_n \to \sigma \text{ in } L^2([0,t] \times \Omega),$$

$$\sigma_n^2 \to \sigma^2 \text{ in } L^1([0,t] \times \Omega),$$

and because a is bounded $a\sigma_n^2 \to a\sigma^2$ in $L^1([0,t] \times \Omega)$. This shows that $\xi_n^2(t,w) - \int_0^t a\sigma_n^2 \, ds$ converges to

$$\xi^2(t,w) - \int_0^t a\sigma^2 \, ds$$

for each t in L^1. Therefore

$$\xi^2(t,w) - \int_0^t a\sigma^2 \, ds$$

is a martingale.

(iv) Let σ be bounded. To show that $\xi \in I[0,\sigma^2]$ it is enough to show that

$$e^{\theta\xi(t,w) - \frac{\theta^2}{2}\int_0^t a\sigma^2 \, ds}$$

is a martingale for each θ, the other conditions being trivially satisfied. Let

$$Z_n(t,w) = e^{\theta\xi_n(t,w) - \frac{\theta^2}{2}\int_0^t a\sigma_n^2 \, ds}$$

We can assume that $|\sigma_n| \leq C$ if $|\sigma| \leq C$ (see the proof of Lemma 3).

$$Z_n = \exp\left[2\theta\xi_n(t,w) - \frac{(2\theta)^2}{2}\int_0^t a\sigma_n^2 \, ds + \theta^2\int_0^t a\sigma_n^2 \, ds\right].$$

Thus

$$(**) \qquad E(Z_n) \leq \text{const } E\left(e^{2\theta\xi_n(t,w) - \frac{(2\theta)^2}{2}\int_0^t a\sigma_n^2 \, ds}\right) = \text{const}$$

since Z_n is a martingale for each θ. A subsequence Z_{n_i} converges to

$$e^{\theta\xi(t,w) - \frac{\theta^2}{2}\int_0^t a\sigma^2 \, ds}$$

almost everywhere (P). This together with (**) ensures uniform integrability of (Z_n) and therefore

$$e^{\theta\xi(t,w) - \frac{\theta^2}{2}\int_0^t a\sigma^2 \, ds}$$

is a martingale. Thus ξ is an Ito process, $\xi \in I[0,a\sigma^2]$.

<u>DEFINITION</u>. With the hypothesis as in the above theorem we define the stochastic integral

$$\xi(t,w) = \int_0^t \sigma(s,w)\, dY(s,w).$$

<u>Exercise</u>. Show that $d(X + Y) = dX + dY$.

<u>REMARK</u>. If σ is bounded, then ξ satis fies the hypothesis of the previous theorem and so one can define the integral of ξ with respect to Y. Further, since ξ itself is Itô, we can also define stochastic integrals with respect to .

<u>Examples</u>. 1. Let $\{\beta(t): t \geq 0\}$ be a Brownian motion; then $\beta(t,w)$ is progressively measurable (being continuous and $\mathcal{F}_t$-measurable). Also,

$$\int_\Omega \int_0^t \beta^2(s)\, ds\, dP = \int_0^t \int_\Omega \beta^2(s)\, dP\, ds = \int_0^t s ds = \frac{t}{2}$$

Hence

$$\int_0^t \beta(s,w)\, d\beta(s,w)$$

is well defined.

2. Similarly $\int_0^t \beta(s/2)\, d\beta(s)$ is well defined.

3. However

$$\int_0^t \beta(2s)\, d\beta(s)$$

is not well defined, the reason being that $\beta(2s)$ is not progressively measurable.

Exercise 5. Show that $\beta(2s)$ is not progressively measurable. (Hint: Try to show that $\beta(2s)$ is not $\mathcal{F}_s$ -measurable for every s. To show this prove that $\mathcal{F}_s \neq \mathcal{F}_{2s}$).

Exercise 6. Show that for a Brownian motion $\beta(t)$, the stochastic integral

$$\int_0^1 \beta(s,.)\, d\beta(s,.)$$

is the same as the left integral

$$L \int_0^1 \beta(s,.)\, d\beta(s,.)$$

defined earlier.

12. CHANGE OF VARIABLE FORMULA

WE SHALL PROVE the

THEOREM. *Let* σ *be any bounded progressively measurable function and* Y *be an Ito process. If* λ *is any progressively measurable function such that*

$$E\left(\int_0^t \lambda^2 ds\right) < \infty,\ \forall\ t,$$

then

$$(*)\qquad \int_0^t \lambda d\xi(s,w) = \int_0^t \lambda(s,w)\,\sigma(s\ ,w)\,dY(s,w),$$

where

$$\xi(t,w) = \int_0^t \sigma(s,w)\,dY(s,w).$$

Proof. Step 1. Let λ and σ be both simple, with λ bounded. By a refinement of the partition, if necessary, we may assume that there exist reals $0 = s_o, s_1, \ldots, s_n, \ldots$ increasing to $+\infty$ such that λ and σ are constant on $[s_j, s_{j+1})$, say $\lambda = \lambda_j(w)$, $\sigma = \sigma_j(w)$, where $\lambda_j(w)$ and $\sigma_j(w)$ are $\mathcal{F}_{s_j}$ -measurable. In this case (*) is a direct consequence of the definition.

Step 2. Let λ be simple and bounded. Let (σ_n) be a sequence of simple bounded functions as in Lemma 3. Put

$$\xi_n(t,w) = \int_0^t \sigma_n(s,w)\,dY(s\ ,w).$$

By Step 1,

$$(**) \qquad \int_0^t \lambda d\xi_n = \int_0^t \lambda\sigma_n \, dY\,(s,w).$$

Since λ is bounded, $\lambda\sigma_n$ converges to $\lambda\sigma$ in $L^2([0,t] \times \Omega)$. Hence, by definition, $\int_0^t \lambda\sigma_n \, dY(s,w)$ converges to $\int_0^t \lambda\sigma dY$ in probability.

Further,

$$\int_0^t \lambda d\xi_n(s,w) = \lambda(s_o,w)[\xi_n(s_1,w) - \xi_n(s_o,w)] + \ldots$$

$$+ \ldots + \lambda(s_k,w)[\xi_n(t,w) - \xi_n(s_{k-1},w)],$$

where $s_o < s_1 < \ldots \quad \ldots$ is a partition for λ, and $\xi_n(t,w)$ converges to $\xi(t,w)$ in probability for every t. Therefore

$$\int_0^t \lambda d\xi_n(s,w)$$

converges in probability to

$$\int_0^t \lambda d\xi(\,s,w).$$

Taking limit as $n \to \infty$ in (**) we get

$$\int_0^t \lambda d\xi(s,w) = \int_0^t \lambda\sigma dY(s,w).$$

Step 3. Let λ be any progressively measurable function with $E(\int_0^t \lambda^2 ds) < \infty$, $\forall\, t$. Let λ_n be a simple approximation to λ as in Lemma 3. Then, by Step 2,

$$(***) \qquad \int_0^t \lambda_n(s,w\,)d\xi(s,w) = \int_0^t \lambda_n(\,s,w)\sigma(s,w)\, dY(s,w).$$

By definition, the left side above converges to

$$\int_0^t \lambda(s,w)d\xi(s,w)$$

in probability. As σ is bounded $\lambda_n\sigma$ converges to $\lambda\sigma$ in $L^2([0,t] \times \Omega)$. Therefore

$$P(\sup_{0 \leq t \leq T} | \int_0^t \lambda_n\sigma \; dY(s,w) - \int_0^t \lambda\sigma \; dy(s,w)| \geq \varepsilon)$$

$$||a||_\infty 1/\varepsilon^2 \; E(\int_0^t (\lambda_n\sigma - \lambda\sigma)^2 \; ds)$$

(see proof of the main theorem leading to the definition of the stochastic integral). Thus

$$\int_0^t \lambda_n\sigma \; dY(s,w)$$

converges to

$$\int_0^t \lambda\sigma \; dY(s,w)$$

in probability. Let n tend to + in (***) to conclude the proof.

13. EXTENSION TO VECTOR-VALUED ITÔ PROCESSES

DEFINITION. Let $(\Omega, \mathcal{F}, P)$ be a probability space and $(\mathcal{F}_t)$ an increasing family of sub σ-algebras of $\mathcal{F}$. Suppose further that

(i) $$a: [0,\infty) \times \Omega \to S_+^d$$

is a probability measurable, bounded function taking values in the class of all symmetric positive semi-definite $d \times d$ matrices, with real entries;

(ii) $$b: [0,\infty) \times \Omega \to \mathbb{R}^d$$

is a progressively measurable, bounded function;

(iii) $$X: [0,\infty) \times \Omega \to \mathbb{R}^d$$

is progressively measurable, right continuous for every w and continuous almost everywhere (P);

(iv) $$Z(t,.) = \exp\left[< \theta, X(t,.) > - \int_0^t < \theta, b(s,.) > ds - \frac{1}{2}\int_0^t < \theta, a(s,.)\theta > ds\right]$$

is a martingale for each $\theta \in \mathbb{R}^d$, where

$$<x,y> = x_1 y_1 + \ldots + x_d y_d, \quad x,y \in R^d.$$

Then X is called an Itô process corresponding to the parameters b and a, and we write $X \in I[b,a]$

NOTE. 1. $Z(t,w)$ is a real valued function.

2. b is progressively measurable if and only if each b_i is progressively measurable.

3. a is progressively measurable if and only if each a_{ij} is so.

Exercise 1. If $X \in I[b,a]$, then s how that

(i) $$X_i \in I[b_i, a_{ii}],$$

(ii) $$Y = \sum_{i=1}^{d} \theta_i X_i \in I[<\theta, b>, <\theta, a\theta>],$$

where

$$\theta = (\theta_1, \ldots, \theta_d).$$

(Hint: (ii)   (i). To prove (ii) appeal to the definition).

REMARK. If X has a multivariate normal distribution with mean μ and covariance (ρ_{ij}), then $Y = <\theta, X>$ has also a normal distribution with mean $<\theta, \mu>$ and variance $<\theta, \rho\theta>$. Note the analogy with the above exercise. This analogy explains why at times b is referred to as the "mean" and **a** as the "covariance".

Exercise 2. If $\{\beta(t): t \geq 0\}$ is a d-dimensional Brownian motion, then $\beta(t,w) \in I[0,I]$ where $I = d \times d$ identity matrix.

As before one can show that $Y(t,.) = X(t,.) - \int_0^t b(s,w)\, ds$ is an Itô process with parameters 0 and a.

DEFINITION. Let X be a d-dimensional Ito process. $\sigma = (\sigma_1, \ldots, \sigma_d)$ a d-dimensional progressively meas urable function such that

$$E(\int_0^t < \sigma(s,.), \sigma(s,.) > ds)$$

is finite or, equivalently,

$$E\left(\int_0^t \sigma_i^2(s,.)ds\right) < \infty, \quad (i = 1,2,\ldots d).$$

Then by definition

$$\int_0^t < \sigma(s,.),\ dX(s,.) > = \sum_{i=1}^d \int_0^t \sigma_i(s,.)\ dX_i(s,.).$$

<u>PROPOSITION</u>. *Let* X *be a* d *-dimensional Itô process* $X \in I[b,a]$ *and let* σ *be progressively measurable and bounded. If*

$$\xi_i(t,.) = \int_0^t \sigma_i dX_i(s,.),$$

then

$$\xi = (\xi_1,\ldots,\xi_d) \in I[B,A],$$

where

$$B = (\sigma_1 b_1,\ldots,\sigma_d b_d) \quad \text{and} \quad A_{ij} = \sigma_i \sigma_j a_{ij}.$$

<u>Proof</u>. (i) Clearly A_{ij} is progressively measurable and bounded. Since $a \in S_+^d$, $A \in S_+^d$.

(ii) Again B is progressi vely measurable and bounded.

(iii) Since σ is bounded, each $\xi_i(t,.)$ is an Itô process; hence ξ is progressively measurable, right continuous, continuous almost everywhere (P). It only remains to verify the martingale condition.

<u>Step 1</u>. Let $\theta = (\theta_1,\ldots,\theta_d) \in \mathbb{R}^d$. By hy pothesis,

$$(*)\ E(\exp[(\theta_1X_1+\ldots+\theta_dX_d)\Big|_s^t \ldots \int_s^t(\theta_1b_1+\ldots+\theta_db_d)du - \frac{1}{2}\int_0^t \sum\theta_i\theta_j a_{ij}ds] \mid \mathcal{F}_s) =$$

Assume that each σ_i is constant on $[s,t]$, $\sigma_i = \sigma_i(w)$ and $\mathcal{F}_s$-measurable. Then (*) remains true if θ_i are replaced by $\theta_i\theta_i(w)$ and since σ_i's are constant over $[s,t]$, we get

$$(**)\qquad E(\exp[\int_0^t \sum_{i=1}^d \theta_i\sigma_i(s,.)\, dX_i(s,.) - \int_0^t \theta_i b_i\ \sigma_i(s\ ,.)ds$$

$$- \frac{1}{2}\int_0^t \sum \theta_i\theta_j\sigma_i(s,.)\sigma_j(s\ ,.)a_{ij}\ ds]|\quad {}_s)$$

$$= \exp[\int_0^s \sum_{i=1}^d \theta_i\sigma_i(s,.)dX_i(s,.) - \int_0^s < \theta,B > du - \frac{1}{2}\int_0^s < \theta,A\theta > du].$$

Step 2. Let each σ_i be a simple function.

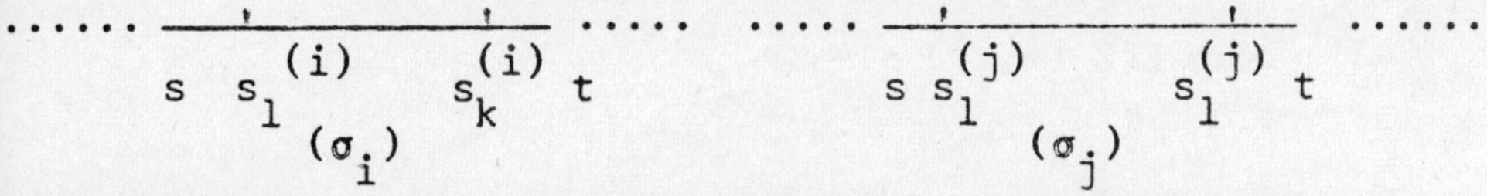

By considering finer partitions we may assume that each σ_i is a step function,

finest partition

i.e. there exist points $s_o, s_1, s_2, \ldots, s_n$, $s = s_o < s_1 < \ldots < s_{n+1} = t$, such that on $[s_j, s_{j+1})$ each σ_i is a constant and $_{s_j}$-measurable. Then (**) holds if we use the fact that if $\mathcal{C}_1 \supset \mathcal{C}_2$.

$$E(E(f|\mathcal{C}_1)|\mathcal{C}_2) = E(f|\mathcal{C}_2).$$

Step 3. Let σ be bounded, $|\sigma| \leq C$. Let $(\sigma^{(n)})$ be a sequence of simple functions approximating σ as in Lemma 3. (**) is true if σ_i

is replaced by $\sigma_i^{(n)}$ for each n. A simple verification shows that the expression $Z_n(t,.)$, in the parenthes is on the left side of (**) with σ_i replaced by $\sigma_i^{(n)}$, converges to

$$Z(t,.) =$$

$$= \operatorname{Exp}\Big(\int_0^t \sum_i \theta_i \sigma_i(s,.)ds - \int_0^t \sum_i \theta_i b_i \sigma_i(s,.)\, ds -$$

$$- \frac{1}{2} \int_0^t \sum_{i,j} \theta_i \theta_j \sigma_i \sigma_j a_{ij}\, ds \Big)$$

as $n \to \infty$ in probability. Since $Z_n(t,.)$ is a martingale and the functions σ_i, σ_j, a are all bounded,

$$\sup_n E(Z_n(t,.)) < \infty.$$

This proves that $Z(t,.)$ is a martingale.

COROLLARY. *With the hypothesis as in the above propos ition define*

$$Z(t) = \int_0^t < \sigma(s,.), dX(s\ ,.) >.$$

Then

$$Z(t,.) \in I[<\sigma,b>, \sigma a \sigma^*]$$

where σ^* *is the transpose of* σ.

Proof. $Z(t,.) = \xi_1(t,.) + \ldots + \xi_d(t,.)$.

DEFINITION. Let $\sigma(s,w) = (\sigma_{ij}(s,w))$ be a $n \times d$ matrix of progressively measurable functions w ith

$$E(\int_0^t \sigma_{ij}^2(s,.)ds) < \infty.$$

If X is a d-dimensional Itô process, we define

$$\left(\int_0^t \sigma(s,.)\, dX(s,.)\right)_i = \sum_{j=1}^{d} \int_0^t \sigma_{ij}(s,.) dX_j(s,.).$$

Exercise 3. Let

$$Z(t,w) = \int_0^t \sigma(s,.)\, dY(s,.),$$

where $Y \in I[0,a]$ is a d-dimensional Itô process and σ is as in the above definition. Show that

$$Z(t,.) \in I[0,\sigma a \sigma^*]$$

is an n-dimensional Ito process, (assume that σ is bounded).

Exercise 4. Verify that

$$E(|Z(t)|^2) = E\left(\int_0^t tr(\sigma a \sigma^*) ds\right).$$

Exercise 5. Do exercise 3 with the assumption that $\sigma a \sigma^*$ is bounded.

Exercise 6. State and prove a change of variable formula for stochastic integrals in the case of several dimensions.

(Hint: For the proof, use the change of variable formula in the one dimensional case and $d(X + Y) = dX + dY$).

14. BROWNIAN MOTION AS A GAUSSIAN PROCESS

SO FAR WE have been considering Brownian motion as a Markov process. We shall now show that Brownian motion can be considered as a Gaussian process.

DEFINITION. Let $X \equiv (X_1,\ldots,X_N)$ be an N-dimensional random variable. It is called an N-variate normal (or Gaussian) distributi[on] with mean $\mu \equiv (\mu_1,\ldots,\mu_N)$ and covariance A if the density function is

$$\frac{1}{(2\pi)^{N/2}} \frac{1}{(\det A)^{1/2}} \exp\left(-\frac{1}{2}[(X-\mu)A^{-1}(X-\mu)^*]\right)$$

where A is an $N \times N$ positive definite symmetric matrix.

NOTE. 1. $E(X_i) = \mu_i$.

2. $\mathrm{Cov}(X_i,X_j) = (A)_{ij}$.

THEOREM. $X \equiv (X_1,\ldots,X_N)$ *is a multivariate normal distribution if and only if for every* $\theta \in \mathbb{R}^N$, $\langle \theta, X \rangle$ *is a one-dimensional Gaussian random variable.*

We omit the proof.

DEFINITION. A stochastic process $\{X_t : t \in I\}$ is called a Gaussian process if $\forall\, t_1,t_2,\ldots,t_N \in I$, $(X_{t_1},\ldots,X_{t_N})$ is an N-variate norma[l] distribution.

Exercise 1. Let $\{X_t : t \geq 0\}$ be a one dimensional Brownian motion. Then show that

a) X_t is a Gaussian process.

(Hint: Use the previous theorem and the fact that increments are independent)

b) $E(X_t) = 0, \forall\ t, E(X(t)X(s)) = s \wedge t.$

Let : $[0,1] \times [0,1] \to \mathbb{R}$ be defined by

$$\rho(s,t) = s \wedge t.$$

Define $K: L^2_{\mathbb{R}}[0,1] \to L^2_{\mathbb{R}}[0,1]$ by

$$K f(s) = \int_0^1 \rho(s,t)\ f(t)\ dt.$$

THEOREM. K *is a symmetric, compact operator. It has only a countable number of eigenvalues and has a comp lete set of eigenvectors.*

We omit the proof.

Exercise 2. Let λ be any eigenvalue of K and f an eigenvector belonging to λ. Show that

(a) $\lambda f'' + f = 0$ with $\lambda f(0) = 0 = \lambda f'(1)$.

(b) Using (a) deduce that the eigenvalues are given by

$\lambda_n = 4/(2n+1)^2\pi^2$ and the corresponding eigenvectors are given by

$f_n = \sqrt{2}\ \text{Sin}\ 1/2[(2n+1)\pi t]$ $n = 0,1,2,\ldots$.

Let $Z_0, Z_1, \ldots, Z_n \ldots$ be identically distributed, independent, normal random variables with mean 0 and variance 1. Then we have

PROPOSITION. $Y(t,w) = \sum_{n=0}^{\infty} Z_n(w)\ f_n(t)\ \sqrt{\lambda_n}$

converges in mean for every real t.

Proof. Let $Y_m(t,w) = \sum_{i=0}^{m} Z_i(w)\, f_i(t)\sqrt{\lambda_i}$. Therefore

$$E\left\{(Y_{n+m}(t,.)-Y_n(t,.))^2\right\} = \sum_{n+1}^{n+m} f_i^2(t)\,\lambda_i,$$

$$E(\|Y_{n+m}(.)-Y_n(.)\|^2 \le \sum_{n+1}^{n+m} \lambda_i \to 0.$$

REMARK. As each $Y_n(t,.)$ is a normal random variable with mean 0 and variance $\sum_{i=0}^{n} \lambda_i\, f_i^2(t)$, $Y(t,.)$ is also a normal random variable with mean zero and variance $\sum_{i=0}^{\infty} \lambda_i f_i^2$. To see this one need only observe that the limit of a sequence of normal random variables is a normal random variable.

THEOREM. (MERCER)

$$\rho(s,t) = \sum_{i=0}^{\infty} \lambda_i f_i(t) f_i(s), (s,t) \in [0,1] \times [0,1].$$

The convergence is uniform.

We omit the proof.

Exercise 3. Using Mercer's theorem show that $\{X_t : 0 \le t \le 1\}$ is a Brownian motion, where

$$X(t,w) = \sum_{n=0}^{\infty} Z_n(w)\, f_n(t)\sqrt{\lambda_n}.$$

This exercise now implies that

$$\int_0^1 X^2(s,w)ds = (L^2 \text{ -norm of } X)^2$$

$$= \sum \lambda_n Z_n^2(w),$$

since $f_n(t)$ are orthonormal. Therefore

$$E(e^{-\lambda\int_0^1 X^2(s,.)ds}) = E(e^{-\lambda\sum_{n=0}^{\infty}\lambda_n Z_n^2(w)}) = \prod_{n=0}^{\infty} E(e^{-\lambda\lambda_n Z_n^2}) \quad \text{(by independence of } Z_n\text{)}$$

$$= \prod_{n=0}^{\infty} E(e^{-\lambda\lambda_n Z_0^2})$$

as $Z_o, Z_n \ldots$ are identically distributed. Therefore

$$E(e^{-\lambda\int_0^1 X^2(s,.)ds}) = \prod_{n=0}^{\infty} 1/\sqrt{(1+2\lambda\lambda_n)}$$

$$= \prod_{n=0}^{\infty} 1/\sqrt{(1+\frac{8\quad 8\lambda}{(2n+1)^2\pi^2})}$$

$$= 1/\sqrt{(\cosh)}\sqrt{(2\lambda)}.$$

<u>APPLICATION</u>. If $F(a) = P(\int_0^1 X^2(s)\,ds < a)$, then

$$\int_0^{\infty} e^{-\lambda a}\,dF(a) = \int_{-\infty}^{\infty} e^{-\lambda a}\,dF(a)$$

$$= E(e^{-\lambda\int_0^1 X^2(s)\,ds}) = 1/\sqrt{(\cosh)}\sqrt{(2\lambda)}.$$

15. EQUIVALENT FORM OF ITÔ PROCESS

LET $(\Omega, \mathcal{F}, P)$ BE A probability space with $(\mathcal{F}_t)_{t \geq 0}$ an increasing family of sub σ-algebras of $\mathcal{F}$ such that $\sigma(\bigcup_{t \geq 0} \mathcal{F}_t) = \mathcal{F}$.

Let

(i) $a: [0,\infty) \times \Omega \to S_d^+$ be a progressively measurable, bounded function taking values in S_d^+, the class of all $d \times d$ positive semidefinite matrices with real entries;

(ii) $b: [0,\infty) \times \Omega \to \mathbb{R}^d$ be a bounded, progressively measurable function;

(iii) $X: [0,\infty) \times \Omega \to \mathbb{R}^d$ be progressively measurable, right continuous and continuous a.s. $\forall\ (s,w) \in [0,\infty) \times \Omega$.

For $(s,w) \in [0,\infty) \times \Omega$ define the operator

$$L_{s,w} = \frac{1}{2} \sum_{i,j=1}^{d} a_{ij}(s,w) \frac{\partial^2}{\partial x_i \partial x_j} + \sum_{j=1}^{d} b_j(s,w) \frac{\partial}{\partial x_j}.$$

For f, u, h belonging to $C_0^\infty(\mathbb{R}^d)$, $C_0^\infty([0,\infty) \times \mathbb{R}^d)$ and $C_b^{1,2}([0,\infty) \times \mathbb{R}^d)$ respectively we define $Y_f(t,w)$, $Z_u(t,w)$, $P_h(t,w)$ as follows:

$$Y_f(t,w) = f(X(t,w)) - \int_0^t (L_{s,w}(f)(X(s,w)) ds,$$

$$Z_u(t,w) = u(t,X(t,w)) - \int_0^t (\frac{\partial u}{\partial s} + L_{s,w} u)(s,X(s,w))ds,$$

$$P_h(t,w) = \exp[h(t,X(t,w)) - \int_0^t (\frac{\partial h}{\partial s} + L_{s,w} h)(s,X(s,w)ds -$$

$$- \frac{1}{2} \int_0^t < a(s,w)\nabla_x h(s,X(s,w)), \nabla_x h(s,X(s,w)) > ds].$$

THEOREM. *The following conditions are equivalent.*

(i) $X_\theta(t,w)=\exp[<\theta,X(t,w)> - \int_0^t <\theta,b(s,w)> ds - \int_0^t <\theta,a(s,w)\theta> ds]$

is a martingale relative to $(\Omega, \mathcal{F}_t, P)$, $\forall\, \theta \in \mathbb{R}^d$.

(ii) $X_\lambda(t,w)$ *is a martingale* $\forall_\lambda$ *in* $\mathbb{R}^d$. *In particular* $X_{i\theta}(t,w)$ *is a martingale* $\forall\, \theta \in \mathbb{R}^d$.

(iii) $Y_f(t,w)$ *is a martingale for every* $f \in C_0^\infty(\mathbb{R}^d)$

(iv) $Z_u(t,w)$ *is a martingale for every* $u \in C_0^\infty([0,\infty) \times \mathbb{R}^d)$.

(v) $P_h(t,w)$ *is a martingale for every* $h \in C_b^{1,2}([0,\infty) \times \mathbb{R}^d)$.

(vi) *The result* (v) *is true for functions* $h \in C^{1,2}([0,\infty) \times \mathbb{R}^d)$ with linear growth, i.e. there exist constants A and B such that $|h(x)| \le A|x| + B$.

The functions $\frac{\partial h}{\partial t}$, $\frac{\partial h}{\partial x_i}$, and $-\frac{\partial^2 h}{\partial x_i \partial x_j}$ which occur under the integral sign in the exponent also grow linearly.

REMARK. The above theorem enables one to replace the martingale condition in the definition of an Itô process by any of the six equivalent conditions given above.

Proof. (i) (ii). $X_\lambda(t,.)$ is $\mathcal{F}_t$-measurable because it is progressively measurable. That $E(|X_\lambda(t,w)|) < \infty$ is a consequence of (i) and the fact that a is bounded.

The function $\lambda \overset{\phi}{\to} \frac{X_\lambda(t,w)}{X_\lambda(s,w)}$ is continuous for fixed $t,s,w,(t > s)$. Morera's theorem shows that ϕ is analytic. Let $A \in \mathcal{F}_s$. Then

$$\int_A \frac{X_\lambda(t,w)}{X_\lambda(s,w)}\, dP(w)$$

is analytic. By hypothesis,

$$\int_A \frac{X_\lambda(t,w)}{X_\lambda(s,w)}\, dP(w) = 1, \ \forall\ \lambda \in \mathbb{R}^d.$$

Thus $\int_A \frac{X_\lambda(t,w)}{X_\lambda(s,w)}\, dP(w) = 1$, $\forall$ complex λ. Therefore $E(X_\lambda(t,w)|\mathcal{F}_s) = X_\lambda(s,w)$, proving (ii). (ii) $\Rightarrow$ (iii). Let

$$A(t,w) = \exp\,[-i\int_0^t <\theta, b(s,w)> ds + \frac{1}{2}\int_0^t <\theta, a(s,w)\theta> ds], \theta \in \mathbb{R}^d.$$

By definition, A is progressively measurable and continuous. Also $|\frac{dA}{dt}(t,w)|$ is bounded on every compact set in $\mathbb{R}$ and the bound is independent of w. Therefore $A(t,w)$ is of bounded variation on every interval $[0,T]$ with the variation $\|A\|_{[0,T]}$ bounded uniformly in w. Let $M(t,w) = X_{i\theta}(t,w)$. Therefore

$$\sup_{0 \leq t \leq T} |M(t,w)| \leq e^{1/2\ T} \sup_{0 \leq t \leq T} |<\theta, a\theta>|.$$

By (ii) $M(t,.)$ is a martingale and since

$$E(\sup_{0 \leq t \leq T} |M(t,w)|\, \|A\|_{[0,T]}(w)) < \infty, \forall\ T,$$

$$M(t,.)\, A(t,.) - \frac{1}{2}\int_0^t M(s,.)\, dA(s,.)$$

is a martingale (for a proof see Appendix), i.e. $Y_f(t,w)$ is a martingale when $f(x) = e^{i<\theta,x>}$.

Let $f \in C_0^\infty(\mathbb{R}^d)$. Then $f \in \mathcal{S}(\mathbb{R}^d)$ the Schwartz-space. Therefore by the Fourier inversion theorem

$$f(x) = \int_{\mathbb{R}^d} \hat{f}(\theta)e^{i<\theta,x>} d\theta.$$

On simplification we get

$$Y_f(t,w) = \int_{\mathbb{R}^d} \hat{f}(\Theta)Y_\Theta(t,w)d\Theta$$

where $Y_\theta \equiv Y_{e^{i<\theta,x>}}$. Clearly $Y_f(t,.)$ is progressively measurable and hence $\mathcal{F}_t$ -measurable.

Using the fact that

$$E(|Y_\theta(t,w)|) \leq 1+t\ d|\theta|\ \|b\|_\infty + \frac{d^2}{2}|\theta|^2\|a\|_\infty,$$

the fact that $\mathcal{S}(\mathbb{R}^d) \subset L^1(\mathbb{R}^d)$ and that $\mathcal{S}(\mathbb{R}^d)$ is closed under multiplication by polynomials, we get $E(|Y_f(t,w)|) < \infty$. An application of Fubini's theorem gives $E(Y_f(t,w)|\mathcal{F}_s) = Y_f(s,w)$, if $t > s$. This proves (iii).

(iii) (iv). Let $u \in C_0([0,\infty) \times \mathbb{R}^d)$.

Clearly $Z_u(t,.)$ is progressively measurable. Since $Z_u(t,w)$ is bounded for every w, $E(|Z_u(t,w)|) < \infty$. Let $t > s$. Then

$$E(Z_u(t,w) - Z_u(s,w)|\mathcal{F}_s) =$$

$$=E(u(t,X(t,w)-u(s,X(s,w)|\mathcal{F}_s)-E(\int_s^t(\frac{\partial u}{\partial\sigma}+L_{\sigma,w}u)(\sigma,X(\sigma,w)\ d\sigma|\mathcal{F}_s)$$

$$=E(u(t,X(t,w)-u(t,X(s,w))|\mathcal{F}_s)+E(u(t,X(s,w)-u(s,X(s,w))|\mathcal{F}_s) -$$

$$- E(\int_s^t(\frac{\partial u}{\partial\sigma} + L_\sigma u_{,w})(\sigma,X(\sigma,w))d\sigma|\mathcal{F}_s)$$

$$= E(\int_s^t (L_{\sigma,w}u)(t,X(\sigma,w))d\sigma|\mathcal{F}_s) +$$

$$+ E\left(\int_s^t \left(\frac{\partial u}{\partial\sigma}(\sigma, X(s,w))\right) d\sigma \,\middle|\, \mathcal{F}_s\right) -$$

$$- E\left(\int_s^t \left(\frac{\partial u}{\partial\sigma} + L_{\sigma}{}^{u}{}_{,w}\right)(\sigma, X(\sigma,w))\,d\sigma \,\middle|\, \mathcal{F}_s\right), \text{ by (iii)}$$

$$= E\left(\int_s^t [L_{\sigma,w}u(t,X(\sigma,w)) - L_{\sigma,w}u(\sigma,X(\sigma,w))]\,d\sigma \,\middle|\, \mathcal{F}_s\right)$$

$$+ E\left(\int_s^t \left[\frac{\partial u}{\partial\sigma}(\sigma,X(s,w)) - \frac{\partial u}{\partial\sigma}(\sigma,X(\sigma,w))\right] d\sigma \,\middle|\, \mathcal{F}_s\right)$$

$$= E\left(\int_s^t [L_{\sigma,w}u(t,X(\sigma,w)) - L_{\sigma,w}u(\sigma,X(\sigma,w))]\,d\sigma \,\middle|\, \mathcal{F}_s\right)$$

$$- E\left(\int_s^t d\sigma \int_s^\sigma L_{\rho,w} \frac{\partial u}{\partial\sigma}(\sigma, X(\rho,w))\,d\rho \,\middle|\, \mathcal{F}_s\right)$$

The last step follows from (iii) (th e fact that $\sigma > s$ gives a minus sign).

$$= E\left(\int_0^t d\sigma \int_\sigma^t \frac{\partial}{\partial\rho} L_{\sigma,w}u(\rho,X(\sigma,w))\,d\rho \,\middle|\, \mathcal{F}_s\right)$$

$$- E\left(\int_s^t d\sigma \int_s^\sigma L_{\rho,w} \frac{\partial u}{\partial\sigma}(\sigma,X(\rho,w))\,d\rho \,\middle|\, \mathcal{F}_s\right)$$

$$= 0$$

(by Fubini). Therefore $Z_u(t,w)$ is a martingale.

Before proving (iv) $\Longrightarrow$ (v) we show that (iv) is true if $u \in C_b^{1,2}([0,\infty) \times \mathbb{R}^d$. Let $u \in C_b^{1,2}$.

(*) Assume that there exists a sequence $(u_n) \in C_0^\infty[[0,\infty) \times \mathbb{R}^d]$ such that

$$u_n \to u, \frac{\partial u_n}{\partial t} \to \frac{\partial u}{\partial t}, \frac{\partial u_n}{\partial x_i} \to \frac{\partial u}{\partial x_i}, \frac{\partial u_n}{\partial x_i \partial x_j} \to \frac{\partial^2 u}{\partial x_i \partial x_j}$$

uniformly on compact sets.

Then $Z_{u_n} \to Z_u$ pointwise and $\sup_n (|Z_{u_n}(t,w)|) < \infty$.

Therefore Z_u is a martingale. Hence it is enough to justify (*).

For every $u \in C_b^{1,2}([0,\infty) \times \mathbb{R}^d)$ we construct a

$u \in C_b^{1,2}((-\infty,\infty) \times \mathbb{R}^d) \equiv C_b^{1,2}(\mathbb{R} \times \mathbb{R}^d)$ as follows.

Put

$$u(t,x) = \begin{cases} u(t,x), & \text{if } t \geq 0, \\ C_1 u(-t,x) + C_2 u(-\frac{t}{2},x), & \text{if } t < 0; \end{cases}$$

matching $\frac{\partial \tilde{u}}{\partial t}, \frac{\partial u}{\partial t}$ at $t = 0$ and $\tilde{u}(t,x)$ and $u(t,x)$ at $t = 0$ yields the desired constants C_1 and C_2. In fact $C_1 = -3$, $C_2 = 4$.

(*) will be proved if we obtain an approximating sequence for $\tilde{u}$.

Let $S: \mathbb{R} \quad \mathbb{R}$ be any C function such that if $|x| \leq 1$,

$$S(x) = \begin{cases} 1, & \text{if } |x| \leq 1, \\ 0, & \text{if } |x| \geq 2. \end{cases}$$

Let $S_n(x) = S(\frac{|x|^2}{n})$ where $|x|^2 = x_1^2 + \dots + x_{d+1}^2$. Put $u_n = S_n \tilde{u}$.

This satisfies (*).

(iv) (v). Let

$$h \in C_b^{1,2}([0,\infty) \times \mathbb{R}^d).$$

Put $u = \exp(h(t,x))$ in (iv) to conclude that

$$M(t,w) = e^{h(t,X(t,w))} - \int_0^t e^{h(s,X(s,w))}[\frac{\partial h}{\partial s} + L_{s,w}h + \frac{1}{2} < \nabla_x h, a\nabla_x h > ds]$$

is a martingale.

Put $A(t,w) = \exp - [\int_0^t \frac{\partial h}{\partial s}(s,w)+L_{s,w}^{-}(s,w)+ \frac{1}{2}< a(s,w)\nabla_x h,\nabla_x h > ds]$.

$A((t,w))$ is progressively measurable, continuous everywhere and $\|A\|_{[0,T]}(w) \leq C_1 e^{C_2 T}$ where C_1 and C_2 are constants. This collows from the fact that $|\frac{dA}{dt}|$ is uniformly bounded in w. Also $\sup_{0 \leq t \leq T} |M(t,w)|$ is uniformly bounded in w. Therefore

$$E(\sup_{0\leq t\leq T} |M(t,w)| \, \|A\|_{[0,T]}(w)) < \infty.$$

Hence $M(t,.)A - \int_0^t M(s,.) \, dA(s,.)$ is a martingale. Now

$$\frac{dA(s,w)}{A(s,w)} = - [\frac{\partial h}{\partial s}(s,w)+L_{s,w} h (s,w)+ \frac{1}{2} < a\nabla_x h,\nabla_x h >]$$

Therefore

$$M(t,w) = e^{h(t,X(t,w))} + \int_0^t e^{h(s,X(s,w))} \frac{dA(s,w)}{A(s,w)}.$$

$$M(t,w)A(t,w) = P_h(t,w)+A(t,w) \int_0^t e^{h(s,X(s,w))} \frac{dA(s,w)}{A(s,w)}$$

$$\int_0^t M(s,.)dA(s,.) = \int_0^t e^{h(s,X(s,w)} dA(s,w)$$

$$+ \int_0^t dA(s,w) \int_0^s e^{h(\sigma,X(\sigma,w))} \frac{dA(\sigma,w)}{A(\sigma,w)}$$

Use Fubini's theorem to evaluate the second integral on the right above and conclude that $P_h(t,w)$ is a martingale.

(vi) $\Longrightarrow$ (i) is clear if we take $h(t,x) = < \theta,x >$. It only remains to prove that (v) $\Rightarrow$ (vi).

(v) $\Rightarrow$ (vi). The technique used to prove this i s an important one and we shall have occasion to use it again.

Step 1. Let $h(t,x) = \theta_1 x_1 + \theta_2 x_2 + \ldots \theta_d x_d = <\theta,x>$ for every $(t,x) \in [0,\infty) \times \mathbb{R}^d$, θ is some fixed element of $\mathbb{R}^d$. Let

$$Z(t) = \exp [<\theta, X_t> - \int_0^t <\theta, b> ds - \frac{1}{2} \int_0^t <\theta, a\theta> ds]$$

We claim that $Z(t,.)$ is a supermartingale.

Let $f: \mathbb{R} \to \mathbb{R}$ be a C^∞ function with compact support such that $f(x) = x$ in $|x| \leq 1/2$ and $|f(x)| \leq 1$, $\forall$ x. Put $f_n(x) = nf(x/n)$. Therefore $|f_n(x)| \leq C|x|$ for some C independent of n and x and $f_n(x)$ converges to x.

Let $h_n(x) = \sum_{i=1}^{d} \theta_i f_n(x_i)$. Then $h_n(x)$ converges to $<\theta,x>$ and $|h_n(x)| \leq C'|x|$ where C is also independent of n and x. By (v),

$$Z_n(t) = \exp[h_n(t,X_t) - \int_0^t (\frac{\partial h_n}{\partial s} + L_{s,w} h_n) ds - \frac{1}{2} \int_0^t < a\nabla_x h_n, \nabla_x h_n > ds]$$

is a martingale. As $h_n(x)$ converges to $<\theta,x>$, $Z_n(t,.)$ converges to $Z(t,.)$ pointwise. Consequently

$$E(Z(t)) = E(\underline{\lim}\, Z_n(t)) \leq \underline{\lim}\, E(Z_n(t)) = 1$$

and $Z(t)$ is a supermartingale.

Step 2. $E(\exp B \sup_{0 \leq s \leq t} |X(s,w)|) < \infty$ for each t and B.

For, let $Y(w) = \sup_{0 \leq s \leq t} |X(s,w)|$, $Y_i(w) = \sup_{0 \leq s \leq t} |X_i(s,w)|$ where where $X = (X_1, \ldots, X_d)$. Clearly $Y \leq Y_1 + \ldots + Y_d$. Therefore

$$E(e^{BY}) \leq E(e^{BY_1} e^{BY_2} \ldots e^{BY_d}).$$

The right hand side above is finite provided $E(e^{BY_i}) < \infty$ for each i as can be seen by the generalised Holder's inequality. Thus to prove the assertion it is enough to show $E(e^{BY_i}) < \infty$ for each $i = 1,2,\ldots d$ with aB' different from B; more specifically for B' bounded.

Put $\theta_2 = 0 = \theta_3 = \ldots = \theta_d$ in Step 1 to get

$$u(t) = \exp[\theta_1 X_1(t) - \int_0^t \theta_1 b_1(s,.)ds - \frac{1}{2}\theta_1^2 \int_0^t a_{11}(s,.)ds]$$

is a supermartingale. Therefore

$$(*) \qquad P(\sup_{0\leq s\leq t} u(s,.) \geq \lambda) \leq \frac{1}{\lambda} E(u(t)) = \frac{1}{\lambda}, \ \forall \lambda > 0.$$

(Refer section on Martingales). Let c be a common bound for both b_1 and a_{11} and let $\theta_1 > 0$. Then (*) reads

$$P(\sup_{0\leq s\leq t} \exp \theta_1 X_1(s) \geq \lambda \exp(\theta_1 ct + \frac{1}{2}\theta_1^2 ct)) \leq \frac{1}{\lambda}.$$

Replacing λ by

$$e^{\lambda\theta_1} e^{-ct\theta_1 - 1/2\, ct\theta_1^2}$$

we get

$$P(\sup_{0\leq s\leq t} \exp \theta_1 X_1(s) \geq \exp \lambda\theta_1) \leq e^{-\lambda\theta_1 + \theta_1 ct + 1/2\, \theta_1^2 ct},$$

i.e. $$P(\sup_{0\leq s\leq t} X_1(s) \geq \lambda) \leq e^{-\lambda\theta_1+\theta_1 ct+1/2\theta_1^2 ct}, \ \forall \theta_1 > 0.$$

Similarly

$$P(\sup_{0\leq s\leq t} - X_1(s) \geq \lambda) \leq e^{-\lambda\theta_1+\theta_1 ct + 1/2\theta_1^2 tc}, \ \forall \theta_1 > 0.$$

As

$$\{Y_1(w) \geq \lambda\} \quad \{\sup_{0 \leq s \leq t} X_1(s) \geq \lambda\} \cup \{\sup_{0 \leq s \leq t} -X_1(s) \geq \lambda\},$$

we get

$$P\{Y_1 \geq \lambda\} \leq 2e^{-\lambda\theta_1 + \theta_1 ct + 1/2\theta_1^2 ct}, \quad \forall\, \theta_1 > 0.$$

Now we get

$$\begin{aligned} E(\exp BY_1) &= \frac{1}{B}\int_0^\infty \exp(Bx)P(Y_1 \geq x)dx \quad (\text{since } Y_1 \geq 0) \\ &\leq \frac{2}{B}\int_0^\infty \exp(Bx - x\theta_1 + \theta_1 ct + \frac{1}{2}\theta_1^2 ct)\, dx \\ &< \infty, \text{ if } B < \theta_1 \end{aligned}$$

This completes the proof of step 2.

<u>Step 3</u>. $Z(t,w)$ is a martingale. For

$$|Z_n(t,w)| = Z_n(t,w)$$

$$= \exp[h_n(X_t) - \int_0^t (\frac{\partial h_n}{\partial s} + L_{s,w}h_n)ds - \frac{1}{2}\int_0^t < a\nabla_x h_n, \nabla_x h_n > ds]$$

$$\leq \exp[h_n(X_t) - \int_0^t L_{s,w}h_n]$$

(since a is positive semidefinite and $\partial h_n/\partial s = 0$).

Therefore $|Z_n(t,w)| \leq A \exp(B \sup_{0 \quad s \quad t} |X(s,w)|)$ (use the fact that $|h_n(x)| \leq C|x|$ and $\frac{\partial h_n}{\partial x_i}, \frac{\partial^2 h_n}{\partial x_i \partial x_j}$ are bounded by the same constant). The result now follows from the dominated convergence theorem and Step 2.

<u>REMARK</u>. In Steps 1, 2 and 3 we have proved that (v) $\Longrightarrow$ (i). The idea of the proof was to express $Z(t,.)$ as a limit of a sequence of

martingales proving first that $Z(t,.)$ is a supermartingale. Using the supermartingale inequality it was then shown that (Z_n) is a uniformly integrable family proving thereby that $Z(t,.)$ is a martingale.

Step 4. Let $h(t,x) \in C^{1,2}([0,\infty) \times \mathbb{R}^d)$ such that $h(t,x)$, $\frac{\partial h}{\partial s}(t,x)$, $\frac{\partial h}{\partial x_i}(t,x)$, $\frac{\partial^2 h}{\partial x_i \partial x_j}(t,x)$ are all dominated by $\alpha|x| + \beta$ for some suitable scalars α and β. Let ϕ_n be a sequence of real valued C^∞ functions defined on $\mathbb{R}^d$ such that

$$\phi_n = \begin{cases} 1 & \text{on } |x| \leq n \\ 0 & \text{on } |x| \geq 2n \end{cases}$$

and suppose there exists a common bound C for

$$\phi_n, \frac{\partial \phi_n}{\partial x_i}, \frac{\partial^2 \phi_n}{\partial x_i \partial x_j} \quad (\forall\, n).$$

Let $h_n(t,x) = h(t,x)\,\phi_n(x)$. By (v) $Z_{h_n}(t,w)$ is a martingale. The conditions on the function h and ϕ_n's show that

$$|Z_{h_n}(t,w)| \leq A \exp (B \sup_{0 \leq s \leq t} |X(s,w)|)$$

where A and B are constants. By Step 2, (Z_{h_n}) are uniformly integrable. Also $Z_{h_n}(t,.)$ converges pointwise to $P_h(t,.)$ (since $h_n \to h$ pointwise). By the dominated convergence theorem $P_h(t,.)$ is a martingale, proving (vi).

16. ITÔ'S FORMULA

MOTIVATION. Let $\beta(t)$ be a one-dimensional Brownian motion. We have seen that the left integral

$$(*) \qquad L[2 \int_0^t \beta(s,.)\, d\beta] = [\beta^2(t,.) - \beta^2(0,.) - t].$$

Formally (*) can be written as

$$d\beta^2(t) = 2\beta(t)\, d\beta(t) + dt.$$

For, on integrating we recover (*).

Newtonian calculus gives the result:

$$df(\beta(t)) = f'(\beta(t))\, d\beta(t) + \frac{1}{2} f''(\beta(t))\, d\beta^2(t) + \ldots$$

for reasonably smooth functions f and β. If β is of bounded variation, only the first term contributes something if we integrate the above equation. This is because $\Sigma d\beta^2 = 0$ for a function of bounded variation. For the Brownian motion we have seen that $\Sigma d\beta^2 \to$ a non zero value, but one can prove that $\Sigma d\beta^3, \ldots\ldots$ converge to 0. We therefore expect the following result to hold:

$$df(\beta(t)) \simeq f'(\beta(t))d\beta(t) + \frac{1}{2} f''(\beta(t))d^2\beta(t).$$

We show that for a one-dimensional Brownian motion

$$\sum(d\beta)^3,\ \sum(d\beta)^4,\ \ldots.$$

all vanish.

$a = t_o \quad t_1 \quad t_2 \quad t_3 \qquad\qquad b = t_{n+1}$

$$E(\sum(d\beta)^3) = E(\sum_{i=0}^{n} [\beta(t_{i+1})-\beta(t_i)]^3) = \sum_{i=0}^{n} E[(\beta(t_{i+1})-\beta(t_i)]^3$$

$$= \sum_{i=1}^{n} 0 = 0,$$

because $\beta(t_{i+1}) - \beta(t_i)$ is a normal random variable with mean zero and variance $t_{i+1} - t_i$. Similarly the higher odd moments vanish. Even moments of a normal random variable with mean 0 and variance σ^2 are connected by the formula

$$\mu_{2k+2} = \sigma^2(2k + 1)\mu_{2k}, \; k > 1.$$

So

$$\sum(d\beta)^4 = \sum_{i=0}^{n} (\beta(t_{i+1}) - \beta(t_i))^4.$$

Therefore

$$E(\sum(d\beta)^4) = \sum_{i=0}^{n} E[(\beta(t_{i+1}) - \beta(t_i))^4]$$

$$= 3\sum_{i=0}^{n} (t_{i+1} - t_i)^2;$$

the right hand side converges to 0 as the mesh of the partition goes to 0. Similarly the higher order even moments vanish.

More generally, if $\beta(t,.)$ is a d -dimensional Brownian motion then we expect

$$df(\beta(t)) \simeq \nabla f(\beta(t)).d\beta(t) + \frac{1}{2}\sum_{i,j} \frac{\partial^2 f}{\partial x_i \partial x_j} d\beta_i d\beta_j .$$

However $\int d\beta_i d\beta_j = 0$ if $i \neq j$ (see exercise below). Therefore

$$df(\beta(t)) \simeq \nabla f(\beta(t)).d\beta(t) + \frac{1}{2} \Delta f(\beta(t))d\beta^2(t).$$

The appearance of Δ on the right side above is related to the heat equation.

Exercise 4. Check that $d\beta_i d\beta_j = \delta_{ij}\ dt$.

(Hint: For $i = j$, the result was proved earlier. If $i \neq j$, consider a partition $0 = t_o < t_1 < \ldots < t_n = t$. Let $\Delta_k\beta_i = \beta_i(t_k) - \beta_i(t_{k-1})$. Then

$$E(\sum_{k=1}^{n} \Delta_k\beta_i\Delta_k\beta_j)^2 = \sum_k E(\Delta_k^2\beta_i\Delta_k^2\beta_j) + 2\sum_{k\neq 1} E[(\Delta_k\beta_i)(\Delta_1\beta_j)];$$

the right side converges to 0 as $n \to \infty$ because $\Delta_k\beta_i$ and $\Delta_\ell\beta_j$ are independent for $k \neq 1$).

Before stating Itô's formula, we prove a few preliminary results.

LEMMA 1. *Let* $X(t,.) \in I[b,0]$ *be a one-dimensional Itô process. Then*

$$X(t,.) - X(0,.) = \int_0^t b(s,.)\,ds \quad \text{a.e.}$$

Proof. $\exp\ [\theta X(t,.) - \theta X(0,.) - \theta\int_0^t b(s,.)\,ds]$ is a martingale for each θ. Therefore

$$E(\exp[\theta(X(t,.)-X(0,.)) - \theta\int_0^t b(s,.)ds]) = \text{constant} = 1,\ \forall\ t.$$

Let

$$W(t,.) = X(t,.) - X(0,.) - \int_0^t b\,(s,.)ds.$$

Then

$$E(\exp\ \theta\ W(t,.)) = \text{Moment generating function of } w = 1,\ \forall\ t.$$

Therefore $\theta(t,.) = 0$ a.e.

REMARK. If $X(t,.) \in I[0,0]$ then $X(t,.) = X(0,.)$ a.e.; i.e. $X(t,.)$ is a trivial process.

We now state a theorem, which is a particular case of the theorem on page 103.

THEOREM. *If* $h \in C^{1,2}([0,\infty) \times \mathbb{R}^d)$ *such that* (i) $|h(x)| \leq A|x| + B$, $\forall\, x \in [0,\infty) \times \mathbb{R}^d$, *for constants* A *and* B (ii) $\frac{\partial h}{\partial t}, \frac{\partial h}{\partial x_i}, \frac{\partial^2 h}{\partial x_i \partial x_j}$ *also grow linearly, then*

$$\exp[h(t,\beta(t,.) - \int_0^t (\frac{\partial h}{\partial s} + \frac{1}{2}\Delta h)(s,\beta(s,.) - \frac{1}{2}\int_0^t |\nabla h|^2 (s,\beta(s,.))ds]$$

is a martingale.

ITO'S FORMULA. *Let* $f \in C_0^{1,2}([0,\infty) \times \mathbb{R}^d)$ *and let* $\beta(t,.)$ *be a d-dimensional Brownian motion. Then*

$$f(t,\beta(t)) - f(0,\beta(0)) = \int_0^t \frac{\partial f}{\partial s}(s,\beta(s,.))\,ds +$$

$$+ \int_0^t < \nabla f(s,\beta(s,.)), d\beta(s,.) > + \frac{1}{2}\int_0^t \Delta f(s,\beta(s,.))ds.$$

where

$$\Delta \equiv \frac{\partial^2}{\partial x_1^2} + \dots + \frac{\partial^2}{\partial x_d^2} .$$

Proof. Step 1. Consider a $(d+1)$-dimensional process defined by

$$X_o(t,.) = f(t,\beta(t,.)),$$

$$X_j(t,.) = \beta_j(t,.).$$

We claim that $X(t,.) \equiv (X_o, X_1, \dots, X_d)$ is a $(d+1)$-dimensional Itô-process with parameters

$$b = [(\frac{\partial f}{\partial s} + \frac{1}{2}\Delta f)(s,\beta(s,.)), \underbrace{0,0,\dots 0}_{d \text{ terms}}]$$

and

$$a = \begin{pmatrix} a_{00} & a_{01} & \cdots & a_{0d} \\ a_{10} & & & \\ \vdots & & I_{d\times d} & \\ a_{d0} & & & \end{pmatrix}$$

where

$$a_{00} = |\nabla_x f|^2 (s,\beta(s,.)),$$

$$a_{0j} = (\frac{\partial}{\partial x_j} f) (s,\beta(s,.)),$$

$$a_{j0} = a_{0j}.$$

For, put $h = \lambda f(t,x) + <\theta, x>$, $x = (x_1,\ldots,x_d) \in \mathbb{R}^d$, in the previous theorem. Then

$$\frac{\partial h}{\partial s} = \lambda\frac{\partial f}{\partial s}, \ \Delta h = \lambda\Delta f, \ \frac{\partial h}{\partial x_j} = \lambda\frac{\partial f}{\partial x_j} + \theta_j.$$

Therefore we see that

$$\exp [\lambda f(t,\beta(t,.)) + <\theta,\beta(t,.)> - \lambda \int_0^t (\frac{\partial f}{\partial s} + \frac{1}{2}\Delta_x f)(s,\beta(s,.)\, ds$$

$$- \frac{1}{2}\lambda^2 \int_0^t |\nabla f|^2(s,\beta(s,.))ds - \frac{1}{2}|\theta|^2 t - \lambda <\theta, \int_0^t \nabla(f(s,\beta(s,.)))ds>]$$

is a martingale.

Consider $(\lambda,\theta)\, a\binom{\lambda}{\theta}$. We have

$$a\begin{pmatrix}\lambda \\ \theta\end{pmatrix} = \begin{pmatrix} a_{00}\lambda + \sum_{j=1}^d a_{0j}\theta_j \\ \rho + \theta \end{pmatrix}, \ \rho = \lambda\begin{pmatrix} a_{10} \\ \vdots \\ a_{d0}\end{pmatrix}.$$

Therefore

$$(\lambda,\theta)a\begin{pmatrix}\lambda\\ \theta\end{pmatrix} = a_{00}\lambda^2 + \lambda\sum_{j=1}^{d} a_{0j}\theta_j + \sum_{j=1}^{d} a_{0j}\theta_j +$$

$$= \lambda^2|\nabla f|^2 + 2\lambda \frac{\partial f}{\partial x_j}\theta_j + |\theta|^2.$$

Thus (*) reads

$$\exp[\lambda f(t,\beta(t,.)+<\theta,\beta(t,.)> - \lambda\int_0^t b_0(s,.)ds - \frac{1}{2}\int_0^t <\alpha,a\alpha> ds]$$

is a martingale where $\alpha = (\lambda,\theta) \in \mathbb{R}^{d+1}$. This proves the claim made above.

<u>Step 2</u>. Define $\sigma(s,.) = (1, -\nabla_x f(s,\beta(s,.)))$ and let

$$Z(t,.) = \int_0^t < \sigma(s,.), dX(s,.) > \text{ where } X = (X_0, X_1, .., X_d)$$

is the (d+1) -dimensional Itô process obtained in Step 1. Since $f \in C_b^{1,2}$, $Z(t,.)$ is an Itô process with parameters $<\sigma,b>$ and $\sigma a\sigma^*$:

$$<\sigma,b> = \frac{\partial f}{\partial s} + \frac{1}{2}\Delta f,$$

$$a\sigma^* = \begin{pmatrix} a_{00} & \rho \\ \rho^* & I \end{pmatrix}, \begin{pmatrix} 1 \\ -\nabla_x f \end{pmatrix}$$

$$= \begin{pmatrix} a_{00} & -<\rho,\nabla_x f> \\ & \\ \rho^* - \nabla_x f & \\ & \end{pmatrix} = \begin{pmatrix} 0 \\ 0 \\ \cdot \\ \cdot \\ 0 \end{pmatrix}$$

Therefore $\sigma a\sigma^* = 0$. Hence by Lemma 1,

$$Z(t,.) - Z(0,.) - \int_0^t <\sigma,b> \;ds = 0 \text{ (a.e.)},$$

$$Z(t,.) = \int_0^t dX_0(s) - \int_0^t < \nabla f(s,\beta(s,.)), d\beta(s,.) >$$

$$= f(t,\beta(t))-f(0,\beta(0))- \int_0^t < \nabla f(s,\beta(s,.)),d\beta(s,.) >$$

Hence $Z(0) = 0$. Thus

$$f(t,\beta(t))-f(0,\beta(0))- \int_0^t < \nabla f(s,\beta(s\ ,.))d\beta(s,.) > -$$

$$- \int_0^t (\frac{\partial f}{\partial s} + \frac{1}{2}\Delta_x f)\ (s,\beta(s,.))\ ds = 0 \quad \text{a.e.}$$

This establishes Itô's formula.

Exercise. (Itô's formula for the general case). Let

$$\phi(t,x) \in C_b^{1,2}([0,\infty) \times \mathbb{R}^d).$$

If $X(t,.)$ is a d-dimensional Itô process corresponding to the parameters b and a, then the following formula holds:

$$\phi(t,X(t,w)) - \phi(0,X(0,w))$$

$$= \int_0^t \frac{\partial\phi}{\partial s}\ (s,X(s,x))ds+ \int_0^t < \nabla_x\phi ,dX > + \frac{1}{2} \int_0^t \sum a_{ij} \frac{\partial^2}{\partial x_i \partial x_j}\ ds.$$

This is also w itten as

$$d\phi(s,X(s,w)) = \phi_s ds + < \nabla_x \phi ,dX > + \frac{1}{2} \sum a_{ij} \frac{\partial^2\phi}{\partial x_i \partial x_j}\ ds.$$

To prove this formula proceed as follows.

(i) Take $h(t,x) = \lambda\phi(t,x) + < \theta ,x >$ in (vi) of the theorem on the equivalence of Itô process to conclude that

$$Y(t,.) = (\phi(t,X(t,.)),X(t,.))$$

is a $(d+1)$-dimensional Ito process with parameters

$$b' = \left(\frac{\partial\phi}{\partial t} + L_{s,w}\phi, b\right)$$

and

$$A = \begin{vmatrix} \langle a\nabla_x\phi, \nabla_x\phi \rangle, & a\nabla_x\phi \\ 1 \times 1 & 1 \times d \\ a\nabla_x\phi & a \\ d \times 1 & d \times d \end{vmatrix}$$

(ii) Let $\sigma(t,x) = (1, -\nabla_x\phi(t,x))$ and

$$Z(t,.) = \int_0^t \langle \sigma(s,X(s,.)), dY(s,.) \rangle .$$

The assumptions on ϕ imply that Z is an Itô process corresponding to

$$(\langle \sigma, b' \rangle, \sigma A \sigma^*) \equiv \left(\frac{\partial\phi}{\partial t} + \frac{1}{2}\sum a_{ij}\frac{\partial^2\phi}{\partial x_i \partial x_j}, 0\right).$$

(iii) Use (ii) to conclude that

$$Z(t,.) = \int_0^t \langle \sigma, b' \rangle \, ds \quad \text{a.e.}$$

This is Itô's formula.

(iv) (Exercise). Verify that Itô's formula agrees with the formula obtained for the case of Brownian motion.

NOTE. Observe that Itô's formula does not depend on b.

Examples 1. Let $\beta(t)$ be a one-dimensional Brownian motion. Then

$$\begin{aligned} d(e^t\phi(\beta(t))) &= e^t d\phi(\beta(t)) + \phi(\beta(t))\, d(e^t) \\ &= e^t\phi'(\beta(t))\, d\beta(t) + \phi(\beta(t))\, e^t dt + \\ &\quad + \frac{1}{2}\phi''(\beta(t))\, e^t\, dt. \end{aligned}$$

2. To evaluate $d\left(e^{\int_0^t V(\beta(s,.))ds} u(t,\beta(t))\right)$

where V is a smooth function, put

$$X_2(t,.) = \int_0^t V(\beta(s,.))ds, \quad b = \begin{pmatrix} 0 \\ V(\quad (t,.)) \end{pmatrix} = \begin{pmatrix} b_1 \\ b_2 \end{pmatrix}$$

$$a = \begin{pmatrix} 1 & 0 \\ 0 & 0 \end{pmatrix}. \quad \text{Let} \quad X_1(t,.) = \beta(t,.), X = (X_1,X_2).$$

Then

$$\exp\left[\theta_1 X_1(t,.) - \theta_2 X_2(t,.) - \theta_1 \int_0^t b_1(X(s,.))ds\right.$$

$$\left. - \theta_2 \int_0^t b_2(X(s,.))ds - \frac{1}{2}\int_0^t < a\theta, \theta > ds\right]$$

$$= \exp\left[\theta_1 X_1(t,.) - \frac{\theta_1^2}{2} t\right];$$

the right side is a martingale. Therefore (X_1,X_2) is a 2-dimensional Itô process with parameters b and a and one can use Itô's formula to write

$$d\left(e^{\int_0^t V(\beta(s,.))ds} u(t,\beta(t))\right) = d\left(e^{X_2(t)} u(t,\beta(t))\right)$$

$$= e^{\int_0^t V(\beta(s,.)} \frac{\partial}{\partial t} u(t,\beta(t))\, dt +$$

$$+ e^{\int_0^t V(\beta(s,.))dt} \frac{\partial}{\partial x} u(t,\beta(t))d\ (t) + e^{\int_0^t V(\beta(s,.))ds} u(t,\beta(t))\, dX_2$$

$$+ \frac{1}{2} e^{\int_0^t V(\beta(s,.))ds} \frac{\partial}{\partial t} u(t,\beta(t))\, dt$$

$$= e^{\int_0^t V(\beta(s,.))dt} \frac{\partial}{\partial t} u(t,\beta(t))dt + \frac{\partial}{\partial t} u(t,\beta(s))d\ (t) + \frac{1}{2}\frac{\partial^2}{\partial x^2} u(t,\beta(t))dt$$

$$+ u(t,\beta(t))\, V(\beta(t,.)dt].$$

3. Let σ_i, f_i, $i = 1,2,..k$ be bounded and progressively measurable relative to a one-dimensional Brownian motion $(\Omega, \mathcal{F}_t, P)$. Write

$$X_i(t,.) = \int_0^t \sigma_i(s,.)\, d\beta(s,.) + \int_0^t f_i(s,.)ds.$$

Then $X_i(t,.)$ is an Itô process with parameters $(\int_0^t f_i(s,.)ds, \sigma_i^2)$ and $(X_1,\ldots,X_k)$ is an Ito process with parameters

$$B = (\int_0^t f_1(s,.)\, ds,\ldots, \int_0^t f_k(s,.)\, ds),$$

$$A = (A_{ij}) \quad \text{where} \quad A_{ij} = \sigma_i \sigma_j \delta_{ij}.$$

If $\phi \equiv \phi(t,X_1(t)\ldots,X_k(t))$, then by Itô's formula

$$d\phi = \frac{\partial\phi}{\partial s} ds + \frac{\partial\phi}{\partial x_1} dX_1 + \ldots + \frac{\partial\phi}{\partial x_k} dX_k$$

$$+ \frac{1}{2} \sum \sigma_i \sigma_j \delta_{ij} \frac{\partial^2\phi}{\partial x_i \partial x_j} ds$$

$$= \frac{\partial\phi}{\partial s} ds + \frac{\partial\phi}{\partial x_1} dX_1 + \ldots + \frac{\partial\phi}{\partial x_k} dX_k$$

$$+ \frac{1}{2} \sum_{i=1}^{k} \sigma_i^2 \frac{\partial\phi}{\partial x_i} ds$$

<u>Exercise</u>. Take $\sigma_1 = 1, \sigma_2 = 0, f_1 = f_2 = 0$ above and verify that if

$$\phi = e^{X_2(t,.)} u(t,\beta(t)),$$

then one gets the result obtained in Example 2 above.

We give below a set of rules which can be used to calculate $d\phi$ in practice, where ϕ is as in Example 3 above.

1. With each $d\beta$ associate a term $\sqrt{(dt)}$

2. If $\phi = \phi(t, X_1, \ldots, X_k)$, formally differentiate ϕ using ordinary calculus retaining terms upto the second order to get

$$(*) \qquad d\phi = \frac{\partial\phi}{\partial t} dt + \frac{\partial\phi}{\partial x_1} dX_1 + \ldots + \frac{\partial\phi}{\partial x_k} X_k + \frac{1}{2} \sum \frac{\partial^2\phi}{\partial x_i \partial x_j} dX_i \, dX_j$$

3. Formally write $dX_i = f_i dt + \sigma_i d\beta_i$, $dX_j = \sigma_j d\beta_j + f_j dt$.

4. Multiply $dX_i dX_j$ and retain only the first order term in dt., For $d\beta_i d\beta_j$ substitute $\delta_{ij}\, dt$. Substitute in (*) to get the desired formula.

<u>ILLUSTRATION OF THE USE OF ITÔ CALCULUS.</u> We refer the reader to the section on Dirichlet problem. There it was shown that

$$u(x) = \int_G u(y)\, \pi(x, dy) = E(u(X(\tau)))$$

satisfies $\Delta u = 0$ in a region G with $u = u(X(\tau))$ on the boundry of G (here τ is the first hitting time).

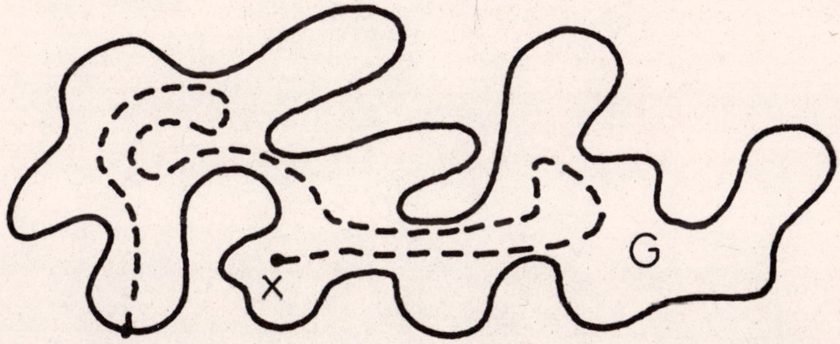

The form of the solution is given directly by Itô's formula (without having recourse to the mean value property). If $u = u(X(t))$ satisfies $\Delta u = 0$ then by Itô's formula

$$du(X(t)) = \langle \nabla u, dX \rangle.$$

Therefore

$$u(X(t)) = u(X(0)) + \int_0^t < \nabla u(\dot{X}(s)), dX(s) >$$

Assuming ∇u to be bounded, we see that $u(X(t))$ is a martingale. By the optional stopping theorem

$$E(u(X(\tau)) = u(X(0)) = u(x).$$

Thus Itô's formula connects solutions of certain differential equations with the hitting probabilities.

17. SOLUTION OF POISSON'S EQUATIONS

LET $X(t,.)$ BE A d-dimensional Brownian motion with $(\Omega, \mathcal{F}_t, P)$ as usual. Let $u(x)\colon \mathbb{R}^d \to \mathbb{R}$ be such that $\frac{1}{2}\Delta u = f$. Assume $u \in C_b^2(\mathbb{R}^d)$. Then $L_{s,w} u \equiv \frac{1}{2}\Delta u = f$ and we know that $u(X(t,.)) - \int_0^t f(X(s,.))\,ds$ is a $(\Omega, \mathcal{F}_t, P)$ - martingale. Suppose now that $u(x)$ is defined only on an open subset $G \subset \mathbb{R}^d$ and $\frac{1}{2}\Delta u = f$ on G. We would like to consider

$$Z(t,.) = u(X(t,.)) - \int_0^t f(X(s,.))ds$$

and ask whether $Z(t,.)$ is still a martingale relative to $(\Omega, \mathcal{F}_t, P)$. Let $\tau(w) = \inf\{t\colon X(t,w) \in \partial G\}$. Put this way, the question is not well-posed because $Z(t,.)$ is defined only upto time $\tau(w)$ for u is not defined outside G. Even if at a time $t > \tau(w)$ $X(t,w) \in G$, one needs to know the values of f for $t > \tau(w)$ to compute the integral.

To answer the question we therefore proceed as follows. Let $A_t = [w\colon \tau(w) > t]$. As t increases, A_t have decreasing measures. We shall give a meaning to the statement '$Z(t,.)$ is a martingale on A_t'. Define

$$\overline{Z}(t,.) = u(X(\tau \wedge t,.)) - \int_0^{\tau \wedge t} f(X(s,.))ds.$$

Therefore

$$\overline{Z}(t,.) = \begin{cases} Z(t), & \text{on } A_t \\ Z(\tau,.), & \text{on } (A_t)^c. \end{cases}$$

Since $Z(t,.)$ is progressively measurable upto time τ, $\bar{Z}(t,.)$ is $\mathcal{F}_t$-measurable.

THEOREM. $\bar{Z}(t,.)$ *is a martingale.*

Proof. Let G_n be a sequence of compact sets increasing to G such that $G_n \subset G^o_{n+1}$. Choose a C^∞ function ϕ_n such that $\phi_n = 1$ on G_n and support $\phi_n \subset G$. Put $u_n = \phi_n u$ and $f_n = \frac{1}{2}\Delta u_n$. Then

$$Z_n(t,.) = u_n(X(t,.)) - \int_0^t f_n(X(s,.))ds$$

is a martingale for each n. Put

$$\tau_n = \inf\{t: X(t,.) \notin G_n\}$$

Then $Z_n(\tau_n \wedge t,.)$ is also a martingale (See exercise below). But

$$Z_n(\tau_n \wedge t) = Z(\tau_n \wedge t).$$

Therefore $M_n(t,.) = Z(\tau_n \wedge t,.)$ is a martingale. Observe that $\tau_n \leq \tau_{n+1}$ and since $G_n \uparrow G$ we have $\tau_n \uparrow \tau$. Therefore $Z(\tau_n \wedge t) \to Z(\tau \wedge t)$ (by continuity); also $|M_n(t,.)| \leq \|u\|_\infty + \|f\|_\infty t$. Therefore $Z(\tau \wedge t) = \bar{Z}(t,.)$ is a martingale.

Exercise. If $M(t,.)$ is a $(\Omega,\mathcal{F}_t,P)$-martingale, show that for any stopping time τ, $M(\tau \wedge t,.)$ is also a martingale relative to $(\mathcal{F}_t)$.

[Hint: One has to show that if $t_2 > t_1$,

$$\int_A M(\tau \wedge t_2,w)dP(w) = \int_A M(\tau \wedge t_1,w)\, dP(w),\ \forall A \in \mathcal{F}_{t_1}.$$

The right side =

$$\int_{A\cap(\tau > t_1)} M(t_1,w)dP(w) + \int_{A\cap(\tau < t_1)} M(\tau\ ,w)dP(w).$$

The left side

$$= \int_{A \cap (\tau > t_1)} M(t_2,w)\, dP(w) + \int_{A \cap (\tau < t_1)} M(\tau,w)dP(w) .$$

Now use optional stopping theorem].

LEMMA. *Let* G *be a bounded region and* τ *be as above. Then* $E_x(\tau) < \infty$, $\forall\, x \in G$, *where* $E_x = E^{P_x}$.

Proof. Without loss of generality we assume that G is a sphere of radius R. The function $u(x) = \frac{R^2 - |x|^2}{d} \geq 0$ and satisfies $\frac{1}{2}\Delta u = -1$ in G. By the previous theorem

$$u(X(\tau \wedge t,.) + \int_0^{\tau \wedge t} ds$$

is a martingale. Therefore

$$E_x(u(X(\tau \wedge t,.))) + E_x(\tau \wedge t) = u(X(0)\ ) = u(x).$$

Therefore $E_x(\tau \wedge t) \leq u(x)$ (since $u \geq 0$). By Fatou's lemma, on letting $t \to \infty$, we obtain $E_x(\tau) \leq u(x) < \infty$. Thus the mere existence of a u satisfying $\frac{1}{2}\Delta u = 1$ helps in concluding that $E_x(\tau) < \infty$.

THEOREM. *Let* $u \in C_b^2(G)$ *and suppose that* u *satisfies*

(*) $$\frac{1}{2}\Delta u = f \quad \textit{in}\ G,$$

$$u = g \quad \textit{on}\ \partial G.$$

Then $u(x) = E_x[g] - E_x[\int_0^\tau f(X(s,.))\, ds]$ *solves* (*).

REMARK. The first part of the solution u(x) is the solution of the homogeneous equation, and the second part accounts for the inhomogeneous term.

Proof. Define $\overline{Z}(t,.) = u(X(\tau \wedge t)) - \int_0^{\tau \wedge t} f(X(s,.))ds$. Then $\overline{Z}$ is a martingale. Also $|\overline{Z}| \leq \|u\|_\infty + \tau \|f\|_\infty$. Therefore, by the previous Lemma, $\overline{Z}(t,.)$ is a uniformly integrable martingale. Therefore we can equate the expectations at time $t = 0$ and at time $t = \infty$ to get

$$u(x) = E_x(g) - E_x[\int_0^{\tau} f(X(s,.))ds].$$

18. THE FEYNMAN-KAC FORMULA

WE NOW CONSIDER the modified heat equation

$$(*) \qquad \frac{\partial u}{\partial t} + \frac{1}{2}\Delta u + V(x)\, u(t,x) = 0, \quad 0 \le t \le T,$$

where $u(T,x) = f(x)$. The Feynman-Kac formula says that the solution for $s \le T$ is given by

$$(**) \qquad u(s,x) = E_{s,x}\left(e^{\int_s^T V(X(s))ds} f(X(T))\right).$$

Observe that the solution at time s dep ends on the expectation with respect to the process starting at time s.

NOTE. (**) is to be understood in the follow ing sense. If (*) admits a smooth solution then it must be given by (**). We shall not go into the conditions under which the solution exists. Let

$$Z(t,.) = u(t,X(t,.))\, e^{\int_s^t V(X(\sigma,.)d\sigma}, \quad t \ge s.$$

By Ito's formula (see Example 2 of section 16), we get

$$Z(t,.) = Z(s,.) + \int_s^t e^{\int_s^\lambda V(X(\sigma,.)d\sigma} < \nabla u(\lambda, X(\lambda)), dX(\lambda)> ,$$

provided that u satisfies (*). Assume tentatively that V and ∇u are bounded and progressively measurable. Then $Z(t,.)$ is a martingale. Therefore

$$E_{s,x}(Z(T,.)) = E_{z,x}(Z(s,.)),$$

or

$$E_{s,x}(u(T,X(T))\, e^{\int_s^T V(X(\sigma,.))d\sigma} = u(s,x).$$

This proves the result.

We shall now remove the condition ∇u is bounded and prove the uniqueness of the solution corresponding to (*) under the assumptic that V is bounded above and

$$|u(t,x)| \leq e^{A+|x|^{\alpha}}, \alpha < 2, \quad \text{on} \quad [s,T).$$

In particular, the Feynman-Kac formula extends the uniqueness theorem for the heat equation to the class of unbounded functions satisfying a growth condition of the form given above.

Let ϕ be a C^{∞} function such that $\phi = 1$ on $|x| \leq R$, and $\phi = 0$ outside $|x| > R+1$. Put $u_R(t,x) = u(t,x)\phi$,

$$Z_R(t,x) = u_R(t,x)\, e^{\int_s^t V(X(\sigma))d\sigma}.$$

By what we have already proved, $Z_R(t,.)$ is a martingale. Let

$$\tau_R(\omega) = \inf\{t: t \geq s\ \omega(t) \in S(0;R) = \{|x| \leq R\}\}.$$

Then $Z_R(t \wedge \tau_R,.))$ is also a martingale, i.e.

$$u_R(t \wedge \tau_R, X(t \wedge \tau_R,.))e^{\int_s^{t \wedge \tau_R} V(X(\sigma))d\sigma}$$

is a martingale. Equating the expectations at time $t = s$ and time $t = T$ and using the fact that

$$u_R(t \wedge \tau_R, X(t \wedge \tau_R,.)) = u(t \wedge \tau_R, X(t \wedge \tau_R,.))),$$

we conclude that

$$u(s,x) = E_{s,x}[u(\tau_R \wedge T, X(\tau_R \wedge T,.))e^{\int_s^{\tau_R \wedge T} V(X(s))ds}]$$

$$= E_{s,x}[\chi_{(\tau_R \wedge T)} f(X(T))e^{\int_s^T V(X(s))ds}] +$$

$$+ E_{s,x}[\chi_{(\tau_R \leq T)} u(\tau_R, X(\tau_R) e^{\int_s^{\tau_R} V(X(s))ds}]$$

Consider the second term on the right:

$$|E_{s,x}[\chi_{(\tau_R \leq T)} u(\tau_R, X(\tau_R)) e^{\int_s^{\tau_R} V(X(s))ds}]|$$

$$\leq A' e^{R^\alpha} P[{}_R \leq T]$$

(where A' is a constant given in terms of the constants A and T and the bound of V)

$$= A' e^{R^\alpha} P[\sup_{s \leq \sigma \leq T} |X(\sigma)| \geq R].$$

$P[\sup_{s \leq \sigma \leq T} |X(\sigma)| \geq R]$ is of the order of $e^{-c(T)R^2}$ and since $\alpha < 2$, the second term on the right side above tends to 0 as $R \to \infty$. Hence, on letting $R \to +\infty$, we get, by the bounded convergence theorem,

$$u(s,x) = E_{s,x}[f(X(T)) e^{\int_s^T V(X(s))ds}]$$

<u>APPLICATION</u>. Let $\beta(t,.)$ be a one-dimensional Brownian motion. Recall (Cf. Reflection principle) that $P\{\sup_{0 \leq s \leq t} |\beta(s)| \leq 1\}$ is of the order of $\frac{4}{\pi} e - \frac{\pi^2 t}{8}$. The **Feynman**-Kac formula will be used to explain the occurance of the factor $\frac{\pi^2}{8}$ in th e exponent. Firstt observe that $\frac{\pi^2}{8} = \frac{\lambda^2}{2}$ where λ is the first positive root of $\text{Cos}\, \lambda = 0$. Let

$$\tau(w) = \inf\{t: |\beta(t)| \geq 1\}.$$

Then

$$P\{\sup_{0 \leq s \leq t} |\beta(s,.)| \leq 1\} = P\{\tau \geq t\}.$$

Let $\phi(x) = E_x[e^{\lambda\tau}]$, $\lambda < 0$. We claim that ϕ satisfies

$$(*)\qquad \begin{aligned} &\frac{1}{2}\phi'' + \lambda\phi = 0, \ |x| < 1, \\ &\phi = 1, \ |x| = 1. \end{aligned}$$

Assume ϕ to be sufficiently smooth. Using Itô's formula we get

$$d(e^{\lambda t}\phi(\beta(t)) = e^{\lambda t}\phi'(\beta(t))d\beta(t)+[\lambda\phi(\beta(t))+ \frac{1}{2}\phi''(\beta(t))]e^{\lambda t}dt.$$

Therefore

$$e^{\lambda t}\phi(\beta(t)) - \phi(\beta(0)) = \int_0^t e^{\lambda s}\phi'(\beta(s))d\beta(s) +$$

$$+ \int_0^t [\lambda\phi(\beta(s))+ \frac{1}{2}\phi''(\beta(s))]e^{\lambda s}ds,$$

i.e.

$$e^{\lambda t}\phi(\beta(t)) - \phi(\beta(0))- \int_0^t[\lambda\phi(\beta(s))+ \frac{1}{2}\phi''(\beta(s))]e^{\lambda s}ds$$

is a martingale. By Doob's optional sampling theorem we can stop this martingale at time τ, i.e.

$$e^{\lambda(t\wedge\tau)}\phi(\beta(t\wedge\tau))-\phi(\beta(0))- \int_0^{t\wedge\tau} [\lambda\phi(\beta(s))+ \frac{1}{2}\phi''(\beta(s))]e^{\lambda s}ds$$

is also a martingale. But for $s \leq t\wedge\tau$,

$$\lambda\phi + \frac{1}{2}\phi'' = 0.$$

Thus we conclude that $\phi(\beta(t\wedge\tau))e^{\lambda(\tau\wedge t)}$ is a martingale. Since $\lambda < 0$ and $\phi(\beta(t\wedge\tau))$ is bounded, this martingale is uniformly integrable. Therefore equating the expectation at $t = 0$ and $t = \infty$ we get (since $\phi(\beta(\tau)) = 1$)

$$\phi(x) = E_x[e^{\lambda\tau}].$$

By uniqueness property this must be the s olution. However (*) has a solution given by

$$(x) = \frac{\text{Cos}\ (\sqrt{(2\lambda x)})}{\text{Cos}\ (\sqrt{(2\lambda)})} .$$

Therefore

$$\text{(1)} \qquad E_0[e^{\lambda\tau}] = \frac{1}{\text{Cos}\ (\sqrt{(2\lambda)})} \quad (\lambda < 0) ,$$

If $F(t) = P(\tau \geq t)$, then

$$\int_0^\infty e^{\lambda t}\, dF(t) = E_0(e^{\lambda\tau}).$$

A theorem on Laplace transforms now tells us that (1) is valid till we cross the first singularity of $\frac{1}{\text{Cos}\ (\sqrt{(2\lambda)})}$. This occurs at $\lambda = \frac{\pi^2}{8}$. By the monotone convergence theorem

$$E_0[e^{\tau\,\pi^2/8}] = +\infty$$

Hence $\int_0^\infty e^{\lambda t}\, dF(t)$ converges for $\lambda < \frac{\pi^2}{8}$ and diverges for $\lambda \geq \frac{\pi^2}{8}$. Thus $\frac{\pi^2}{8}$ is the supremum of λ for which $\int_0^\infty e^{\lambda t} dF(t)$ converges, i.e. $\sup\ [\lambda\colon E_0(e^{\lambda\tau})]$ exists, i.e. the decay rate is connected to the existence or the non existence of the solution of the system (*). This is a general feature and prevails even in higher dimensions.

19. AN APPLICATION OF THE FEYNMAN-KAC FORMULA. THE ARC SINE LAW.

LET $\beta(t,.)$ BE THE one-dimensional Brownian motion with $\beta(0,.) = 0$. Define

$$\xi_t(w) = \frac{1}{t}\int_0^t \chi_{[0,\infty)}(\beta(s,w))ds.$$

$\xi_t(w)$ is a random variable and denotes the fraction of the time that a Brownian particle stays above the x-axis during the time interval $[0,t]$. We shall calculate

$$P[w: \xi_t(w) \leq a] = F_t(a)$$

BROWNIAN SCALING. Let $X_t(s) = \frac{1}{\sqrt{t}}\beta(ts)$. Then X_t is also a Brownian motion with same distribution as that of $\beta(s)$. We can write

$$\xi_t(w) = \int_0^1 \chi_{[0,\infty)}(X_t(s,w))ds.$$

Then $\xi_t(w)$ = time spent above the x-axis by the Brownian motion $X_t(s)$ in $[0,1]$. Hence $F_t(a)$ is independent of t and is therefore denoted by $F(a)$. Suppose we succeed in solving for $F(a)$; if, now,

$$\xi_t^*(w) = \int_0^t \chi_{[0,\infty)}(\beta(s))ds = t\xi_t,$$

then the amount of time $\beta(s,.) > 0$ in $[0,t]$ is t (amount of time $X_t(s) > 0$ in $[0,1]$). Hence we can solve for $P[\xi_t^*(w) \leq a] = G_t(a)$. Clearly the solution of G_t is given by

$$G_t(a) = F(a/t).$$

It is clear that

$$F(a) = \begin{cases} 0 & \text{if } a \le 0, \\ 1 & \text{if } a \ge 1. \end{cases}$$

Hence it is enough to solve for $F(a)$ in $0 \le a \le 1$. Let

$$u_\lambda(t,x) = E_x[e^{-(\lambda \int_0^t \chi_{[0,\infty)}(\beta(s,w))ds}]$$

Then

$$u_1(t,0) = E[e^{-\xi(w))}] = \int_0^1 e^{-tx}\, dF(x).$$

Also note that $u_\lambda(t,x)$ is bounded by 1, if $\lambda \ge 0$. By the Feynman-Kac formula (appropriately modified in case $\frac{1}{2}\Delta$ is replaced by $-\frac{1}{2}\Delta$) $u_1(t,x)$ satisfies

$$(*) \quad \begin{aligned} \frac{\partial u}{\partial t} &= \frac{1}{2}\frac{\partial^2 u}{\partial x^2} - u, \quad x > 0, \\ &= \frac{1}{2}\frac{\partial^2 u}{\partial x^2} \quad\quad , \quad x < 0, \end{aligned}$$

and $u(0,x) = 1$. Let

$$\phi_\alpha(x) = \alpha \int_0^\infty u(t,x)e^{-\alpha t}\, dt, \ \alpha > 0, \quad \text{where} \quad u = u_1,$$

$$= \int_0^\infty u(t,x)\ (-de^{-\alpha t}).$$

A simple integration by parts together with (*) gives the following system of ordinary differential equations for ϕ_α:

$$-\frac{1}{2}\phi''_\alpha + (\alpha+1)\phi_\alpha = \alpha, \quad x > 0,$$
$$-\frac{1}{2}\phi''_\alpha + \alpha\phi_\alpha = \alpha \quad , \quad x < 0.$$

These have a solution

$$\phi_\alpha(x) = \frac{\alpha}{\alpha+1} + A\, e^{x\sqrt{(2(\alpha+1))}} + B\, e^{-x\sqrt{(2(\alpha+1))}},\ x > 0,$$

$$= 1 + C\, e^{x\sqrt{(2\alpha)}} + D\, e^{-x\sqrt{(2\alpha)}},\quad x < 0.$$

However u is bounded by 1 (see definition of $u_\lambda(t,x)$). Therefore ϕ_α is also bounded by 1. This forces $A = D = 0$. We demand that ϕ_α and $\frac{d\phi_\alpha}{dx}$ should match at $x = 0$. (Some justification for this will be given later on).

$$\underset{x \to 0+}{\mathrm{Lt}}\ \phi_\alpha(x) = \underset{x \to 0-}{\mathrm{Lt}}\ \phi_\alpha(x)$$

gives

$$\frac{\alpha}{1+\alpha} + B = C + 1.$$

Similarly we get $-B\sqrt{(2(\alpha+1))} = C\sqrt{(2\alpha)}$ by matching $\frac{d\phi_\alpha}{dx}$. Solving for B and C we get

$$B = \frac{\sqrt{\alpha}}{(1+\alpha)\,(\sqrt{\alpha} + \sqrt{(\alpha+1)})},$$

$$C = \frac{1}{\sqrt{(1+\alpha)}\,(\sqrt{\alpha} + \sqrt{(\alpha+1)})}.$$

Therefore

$$\phi_\alpha(0) = \frac{\alpha}{\alpha+1} + B = \frac{\sqrt{\alpha}}{\sqrt{(\alpha+1)}},$$

i.e.

$$\int_0^\infty E[e^{-t\xi_t}\ \alpha e^{-\alpha t}]\, dt = \frac{\sqrt{\alpha}}{\sqrt{(\alpha+1)}}.$$

Using Fubini's theorem this gives

$$E[\frac{\alpha}{\alpha+\xi}] = \sqrt{(\frac{\alpha}{\alpha+1})},$$

or

$$E\left[\frac{1}{1+\xi/a}\right] = \sqrt{\left(\frac{1}{1+(1/\alpha)}\right)}, \text{ i.e. } \int_0^1 \frac{dF(x)}{1+\gamma x} = \frac{1}{\sqrt{(1+\gamma)}} .$$

This can be inverted to get

$$dF(x) = \frac{2}{\pi}\frac{dx}{\sqrt{(x(1-x))}} .$$

(Refer tables on transforms or check directly that

$$\frac{2}{\pi}\int_0^1 \frac{1}{1+\beta x}\frac{dx}{\sqrt{(x(1-x))}} = \frac{1}{\sqrt{(1+\beta)}}$$

by expanding the left side in powers of β). Therefore

$$F(a) = \frac{2}{\pi} \arcsin (\sqrt{a}), \quad 0 \leq a \leq 1.$$

Hence $G_t(a) = \frac{2}{\pi} \arcsin (\sqrt{(\frac{a}{t})}),\ 0 \leq a \leq t,$ i.e.

$$P[\xi_t \leq a] = \frac{2}{\pi} \arcsin (\sqrt{(\frac{a}{t})}),\ 0 \leq a \leq t.$$

This result goes by the name of *arc sine law* for obvious reasons.

We now give some justification regarding the matching conditions used above.

The equation we solved was

$$\alpha\phi - \frac{1}{2}\phi'' + V\phi = f$$

where ϕ was bounded $V \geq 0$. Suppose we formally use Itô's formula to calculate

$$d(\phi(\beta(t)e^{-\int_0^t (\alpha+V)(\beta(s)ds})$$

$$= e^{-\int_0^t (\alpha+V)(\beta(s,.))ds}[-f(\beta(s,.))dt + \frac{d\phi}{dX}(\beta(t))d\beta(t)]$$

(see Example 2 of Itô's formula). Therefore

$$(Z(t,.)=\phi(\beta(t,.))e^{-\int_0^t(\alpha+V)(\beta(s,.))ds} + \int_0^t f\beta(s,.)\exp(-\int_0^s(\alpha+V)d\sigma)ds$$

is a martingale. Since ϕ, f are bounded and $V \geq 0$,

$$|Z(t,.)| \leq \|\phi\|_\infty + \|f\|_\infty \int_0^\infty e^{-\alpha s} ds \leq \|\phi\|_\infty + C\|f\|_\infty .$$

Therefore $Z(t,.)$ is uniformly integrable. Equating the expectations at time 0 and ∞ gives

$$(*) \qquad \phi(0) = E_0 \int_0^\infty [f(\beta(s,.))e^{-\alpha s - \int_0^s V(\beta(\sigma))d\sigma}]ds.$$

This is exactly the form obtained by solving the differential equations. In order to use Itô's formula one has to justify it. If we show that Itô's formula is valid for functions having a discontinuity in the second derivative, (*) will be a legitimate solution and in general there is no reason why the second derivatives (or higher derivatives) should be matched. This partially explains the need for matching ϕ and $\frac{d\phi}{dx}$ only.

PROPOSITION. *Let* $\beta(t,.)$ *denote a one-dimensional Brownian motion. Suppose* $\phi \in C_b^1$ *and satisfies*

$$\alpha\phi - \frac{1}{2}\phi'' + V\phi = f,$$

Then

$$\phi(\beta(t)) - \int_0^t f(\beta(s))ds$$

is a martingale.

Proof. Let $(\phi_\varepsilon) \in C_b^2$ such that $\alpha\phi_\varepsilon - \frac{1}{2}\phi_\varepsilon'' + V\phi_\varepsilon = f_\varepsilon$ and such that (i) ϕ_ε converges to ϕ uniformly on compact sets, (ii) ϕ_ε' converges to ϕ' uniformly on compact sets, (iii) ϕ_ε'' converges pointwise to ϕ'' except at 0. We may suppose that the convergence is bounded.

CLAIM. $\int_0^t f_\varepsilon(\beta(s))ds$ *converges to* $\int_0^t f(\beta(s))ds$ *a.e.* As $f_\varepsilon(\beta(s))$ converges to $f(\beta(s))$ except wh en $\beta(s) = 0$, it is enough to prove that

(*) P[w: Lebesgue measure $(s: \beta(s) = 0) > 0] = 0$.

Let $\chi_{\{0\}}$ denote the indicator function of $\{0\}$. Then

$$E \int_0^t \chi_{\{0\}} (\beta(s))ds = \int_0^t E\, \chi_{\{0\}} (\beta(s))ds = 0.$$

Thus (*) holds and establishes the claim. Now

$$\phi_\varepsilon(\beta(t)) - \int_0^t f_\varepsilon(\beta(s))ds$$

is a uniformly bounded martingale converging to

$$\phi(\beta(t)) - \int_0^t f(\beta(s))ds.$$

Therefore

$$\phi(\beta(t)) - \int_0^t f(\beta(s))\, ds$$

is a martingale.

20. BROWNIAN MOTION WITH DRIFT

LET $\Omega = C[0,\infty; \mathbb{R}^d]$, $\mathcal{F}$ = BOREL σ -FIELD of Ω, $\{X(t,.)\} \equiv$ Brownian motion, $\mathcal{F}_t = \sigma[X(s,.): 0 \leq s \leq t]$, $P_x \equiv$ probability measure on Ω corresponding to the Brownian motion starting at time 0 at x. $\mathcal{F} = \sigma(\bigcup_{t \geq 0} \mathcal{F}_t)$. Let $b: \mathbb{R}^d \to \mathbb{R}^d$ be any bounded measurable function. Then the map $(s,w) \mapsto b(w(s))$ is p rogressively measurable and

$$Z(t,.) = \exp[\int_0^t < b(X(s,.)),dX(s,.) > - \frac{1}{2} \int_0^t |b(X(s,.))|^2 ds]$$

is a martingale relative to $(\Omega, \mathcal{F}_t, P_x)$. Define Q_x^t on $\mathcal{F}_t$ by

$$Q_x^t(A) = \int_A Z(t,.) \ dP_x,$$

i.e. $Z(t,.)$ is the Radon-Nikodym derivative of Q_x^t with respect to P_x on $\mathcal{F}_t$.

PROPOSITION. (i) Q_x^t *is a probability measure.*

(ii) $\{Q_x^t: t \geq 0\}$ *is a consistent family on* $\bigcup_{t \geq 0} \mathcal{F}_t$, *i.e. if* $A \in \mathcal{F}_{t_1}$ *and* $t_2 \geq t_1$ *then* $Q_x^{t_1}(A) = Q_x^{t_2}(A)$.

Proof. Q_x^t being an indefinite integral, is a measure. Since $Z(t,.) \geq 0$, Q_x^t is a positive measure. $Q_x^t(\Omega) = E_x(Z(t,.))=E_x(Z(0,.))=1$. This proves (i). (ii) follows from the fact that $Z(t,.)$ is a martingale.

If $A \in \mathcal{F}_t$, we define

$$Q_x(A) = Q_x^t(A).$$

The above proposition shows that Q_x. is well defined and since $(\mathcal{F}_t)$ is an increasing family, Q_x is finitely additive on $\bigcup_{t \geq 0} \mathcal{F}_t$.

Exercise. Show that Q_x is countably additive on the algebra $\bigcup_{t \geq 0} \mathcal{F}_t$.

Then Q_x extends as a measure to $\mathcal{F} = \sigma(\bigcup_{t \geq 0} \mathcal{F}_t)$. Thus we get a family of measures $\{Q_x : x \in \mathbb{R}^d\}$ defined on $(\Omega, \mathcal{F})$.

PROPOSITION. *If* $s < t$ *then*

$$Q_x(X_t \in A | \mathcal{F}_s) = Q_{X(s)}(X(t-s) \in A) \quad a.e.$$

DEFINITION. If a family of measures $\{Q_x\}$ satisfies the above property it is called a homogeneous Markov family.

Proof. Let $B \in \mathcal{F}_s$. Therefore $B \cap X_t^{-1}(A) \in \mathcal{F}_t$ and by definition,

$$Q_x((X(t) \in A) \cap B)) = \int_{B \cap X_t^{-1}(A)} Z(t,.)dP_x$$

$$= E^{P_x}(Z(t,.)\, \chi_B^{(w)}\, \chi_A(X(t,.)))$$

$$= E^{P_x}(E^{P_x}(\frac{Z(t,.)}{Z(s,.)}\, Z(s,.)\, \chi_B^{(w)}\, \chi_A(X(t,.))| \mathcal{F}_s)$$

$$= E^{P_x}([(\chi_B Z(s,.))E^{P_x}(\frac{Z(t,.)}{Z(s,.)}\, \chi_A(X(t,.))]| \mathcal{F}_s)$$

(since $B \in \mathcal{F}_s$ and $Z(s,.)$ is $\mathcal{F}_s$ -measurable)

$$= E^{Q_x}[\chi_B E^{P_x}(\frac{Z(t,.)}{Z(s,.)}\, \chi_A(X(t,.))| \mathcal{F}_s)] \ldots \qquad (1)$$

$$= E^{Q_x}[\chi_B E^{P_x}(\exp[\int_s^t < b, dX > - \frac{1}{2}\int_0^t |b|^2\,]\chi_A(X(t,.))| \mathcal{F}_s)]$$

$$= E^{Q_x}[\chi_B E^{P_{X(s)}}(\exp [\int_0^{t-s} < b,dX > \frac{1}{2} \int_0^{t-s} |b|^2]\chi_A(X(t-s))]$$

(by Markov property of Brownian motion)

$$= E^{Q_x}(\chi_B E^{Q^{t-s}_{X(s)}}(\chi_A(X(t-s))) \quad (\text{since } \frac{dQ^{t-s}_{X(s)}}{dP_{X(s)}} = Z(t-s,.)$$

$$= E^{Q_x}(\chi_B E^{Q_{X(s)}} \chi_A(X(t-s,.))$$

The result follows from definition.

Let $b: [0,\infty) \times \mathbb{R}^d \to \mathbb{R}^d$ be a bounded measurable function, $P_{s,x}$ the probability measure corresponding to the Brownian motion starting at time s at the point x. Define, for $t \geq s$,

$$Z_{s,t}(w) = \exp [\int_s^t < b(\sigma,X(\sigma,w)),dX(\sigma,w) >$$

$$- \frac{1}{2} \int_s^t |b(\sigma,X(\sigma,w))|^2 d\sigma]$$

Exercise (i) $Z_{s,t}$ is a martingale relative to $(\mathcal{F}^s_t, P_{s,x})$.

(ii) Define $Q^t_{s,x}$ by $Q^t_{s,x}(A) = \int_A Z_{s,t} dP_{s,x}$, $\forall A \in \mathcal{F}^s_t$. Show that $Q^t_{s,x}$ is a probability measure on $\mathcal{F}^s_t$.

(iii) $Q^t_{s,x}$ is a consistent family.

(iv) $Q_{s,x}$ defined on $\bigcup_{t \geq s} \mathcal{F}^s_t$ by $Q_{s,x}|\mathcal{F}^s_t = Q^t_{s,x}$ is a finitely additive set function which is countably additive.

(v) The family $\{Q_{s,x}: 0 \leq s < \infty, x \in \mathbb{R}^d\}$ is an inhomogeneou Markov family, i.e.

$$Q_{s,x}(X(t,.) \in A|\mathcal{F}^s_\sigma) = Q_{\sigma,X(\sigma,.)}(X(t,.) \in A) \ ,\forall\ s < \sigma < t, A \in \mathcal{F}^s_t .$$

[Hint: Repeat the arguments of the previous section with obvious modifications].

PROPOSITION. *Let* τ *be a stopping time,* $\tau \geq s$. *Then*

$$Q_{s,x}[X(t,.) \in A | \mathcal{F}_{\tau}^{s}] = Q_{\tau,X_{\tau}(.)}(X(t,.) \in A) \text{ on } \tau(w) < t,$$

$$= \chi_A(X(t,.)) \quad \text{on} \quad \tau(w) \geq t.$$

Proof. Let $B \in \mathcal{F}_{\tau}^{s}$ and $B \subset \{\tau < t\}$ so that $B \in \mathcal{F}_{t}^{s}$.

$$E^{Q_{s,x}}(\chi_B \chi_A(X_t)) = E^{P_{s,x}}(Z_{s,t}\chi_B\chi_A(X_t))$$

$$= E^{P_{s,x}}[E^{P_{s,x}}(Z_{\tau,t}Z_{s,\tau}\chi_B\chi_A(X_t)|\mathcal{F}_{\tau}^{s})]$$

(since Z satisfies the multiplicative property)

$$= E^{P_{s,x}}(Z_{s,\tau}\chi_B E^{P_{s,x}}(Z_{\tau,t}\chi_A(X_t)|\mathcal{F}_{\tau}^{s})]$$

(since $Z_{s,\tau}$ is $\mathcal{F}_{\tau}^{s}$- measurable)

(*) $$= E^{P_{s,x}}[Z_{s,\tau}\chi_B E^{P_{\tau,X_\tau}}(Z_{\tau,t}\chi_A(X_t))]$$

(by strong Markov property).

Now

$$\left.\frac{dQ_{s,x}}{dP_{s,x}}\right|_{\mathcal{F}_{t}^{s}} = Z_{s,t}.$$

so that the optional stopping theorem,

$$\left.\frac{dQ_{s,x}}{dP_{s,x}}\right|_{\mathcal{F}_{\tau}^{s}} = Z_{s,} \quad \text{on} \quad \{\tau < t\}, \forall x.$$

Putting this in (*) we get

$$E^{Q_{s,x}}[\chi_B \chi_A(X_t)] = E^{P_{s,x}}[Z_{s,\tau} \chi_B E^{Q_{\tau,X_\tau}}(\chi_A X_t)].$$

Observe that

$$\chi_B E^{Q_{\tau,X_\tau}}(\chi_A(X_t))$$

is $\mathcal{F}_\tau^s$ -measurable to conclude the first part of the proof. For part (ii) observe that

$$X_t^{-1}(A) \cap \{\tau \geq t\} \cap \{\tau \leq k\}$$

is in $\mathcal{F}_k^s$ if $k > s$, so that

$$X_t^{-1}(A) \cap \{\tau \geq t\} \in \mathcal{F}_\tau^s .$$

Therefore

$$E^{Q_{s,x}}(X_t \in A \cap (\tau \geq t)) | \mathcal{F}_\tau^s) = \chi_A(X_t)\chi_{\{\tau \geq t\}},$$

or

$$E^{Q_{s,x}}[(X_t \in A) | \mathcal{F}_\tau^s] = \chi_A(X_t) \quad \text{if} \quad \tau \geq t.$$

<u>PROPOSITION</u>. *Let* $b: [0,\infty) \times \mathbb{R}^d \to \mathbb{R}^d$ *bbe a bounded measurable function,* $f: \mathbb{R}^d \to \mathbb{R}$ *any continuous bounded function. If*

$$\frac{\partial u}{\partial s} + \frac{1}{2} \Delta u + < b(s,x), \Delta u > = 0, \quad 0 \leq s \leq t,$$

$$u(t,x) = f(x)$$

has a solution u, *then*

$$u(s,x) = \int_\Omega f(X_t) dQ_{s,x}.$$

REMARK. b is called the *drift*. If $b = 0$ and $s = 0$ then we recover the result obtained earlier. With the presence of the drift term, the result is the same except that instead of $P_{s,x}$ one has to use $Q_{s,x}$ to evaluate the expectation.

Proof. Let

$$Y(\sigma,.) = \int_s^\sigma < b(\theta,X(\theta,.)),dX(\theta,.) > - \frac{1}{2}\int_s^\sigma |b(\theta,X(\theta,.)|^2 d\theta .$$

Step 1. $(X(\sigma,.) - X(s,.),\ Y(\sigma,.))$ is a $(d+1)$-dimensional Itô process with parameters

$$(\underbrace{0,0,\ldots,0}_{d \text{ terms}} ,\ -\frac{1}{2}|b(\sigma,X(\sigma,.))|^2) \text{ and}$$

$$a = \begin{pmatrix} I_{d\times d} & b_{d\times 1} \\ b^*_{1\times d} & b \end{pmatrix}$$

Let $\lambda = (\lambda_1,\ldots,\lambda_d)$. We have to s how that

$$\exp [\mu Y + \frac{\mu}{2}\int_s^\sigma |b(\sigma,X(\sigma,.)|^2 d\sigma + < \lambda,X(\sigma,.) - X(s,.) > -$$

$$- \frac{1}{2}\int_s^\sigma < \lambda,\lambda > + 2\mu < \lambda,b + \mu^2|b(\sigma,.)|^2 d\sigma]$$

is a martingale, i.e. that

$$\exp [<\lambda,X(\sigma,.) - X(s,.) > + \mu \int_s^\sigma < b,dX > - \frac{1}{2}\int_s^\sigma |\lambda + b\mu|^2 d\rho].$$

is a martingale; in other words that

$$\exp\left[\int_s^\sigma < \lambda + b\mu, dX > - \frac{1}{2}\int_s^\sigma |\lambda + b\mu|^2 d\rho\right]$$

is a martingale. But this is obvious because

$$Z(\sigma,.) = \int_s^\sigma < \lambda + \mu b, dX >$$

is a stochastic integral and hence an Itô proces s with parameters $(0, |\lambda + \mu b|^2)$. (Refer to the section on vector-valued Itô process).

Step 2. Put $\phi(\sigma, X(\sigma,.), Y(\sigma,.)) = u(\sigma, X(\sigma,.))\, e^{Y(\sigma,.)}$. By Itô formula,

$$d\phi = e^Y \frac{\partial u}{\partial t} dt + e^Y < \nabla u, dX > + u e^Y dY + \frac{1}{2}\sum a_{ij} \frac{\partial^2 \phi}{\partial z_i \partial z_j},$$

where $z = (x,y)$, or

$$d\phi = e^Y\left[\frac{\partial u}{\partial t} dt + < \nabla u, dX > + u < b, dX > - \frac{u}{2}|b|^2 dt + \frac{1}{2}\Delta u dt + < b, \nabla u > dt + \frac{1}{2} u|b|^2 dt\right]$$

$$= e^Y[< \nabla u, dX > + u < b, dX >].$$

Therefore ϕ is an Itô process and hence a martingale. Therefore

$$E(\phi(t,.)) = E(\phi(s,.))$$

$$u(s,x) = E^{P_{s,x}}\left[(f(X(t))e^{\int_s^t < b, dX > - \frac{1}{2}\int_0^t |b|^2 d\theta}\right]$$

$$= E^{Q_{s,x}}[f(X(t))],$$

which proves the theorem.

Alternate Proof.

Exercise. Let $Y(\sigma,.)$ be progressively measurable for $\sigma \geq s$. Then $Y(\sigma,.)$ is a martingale relative to $(Q_{s,x},\mathcal{F}_t^s)$ if and only if $Y(\sigma)Z_{s,\sigma}$ is a martingale relative to $(P_{s,x},\mathcal{F}_t^s)$.

Now for any function θ which is progressively measurable and bounded,

$$\exp\left[\int_s^t \langle \theta, dX \rangle - \frac{1}{2}\int_s^t |\theta|^2 d\sigma\right]$$

is a martingale relative to $(\Omega,\mathcal{F}_t^s,P_{s,x})$. In particular let θ be replaced by $\theta + b(\sigma,w(\sigma))$. After s ome rearrangement one finds that X_t is an Itô process with parameters b,I relative to $Q_{s,x}$. Therefore

$$u(t,X_t) - \int_s^t \left(\frac{\partial u}{\partial \sigma} + \langle b,\nabla u \rangle + \frac{1}{2}\Delta u\right)d\sigma$$

is a martingale relative to $Q_{s,x}$. But

$$\frac{\partial u}{\partial \sigma} + \langle b, \nabla u \rangle + \frac{1}{2}\Delta u = 0.$$

Therefore

$$E^{Q_{s,x}}(u(t,X(t)) = u(s,X).$$

We have defined $Q_{s,x}$ by using the notion of the Radon-Nikodym derivative. We give one more relation between P and Q.

THEOREM. *Let* $T: C([s,\infty), \mathbb{R}^d) \to C([s,\infty), \mathbb{R}^d)$ *be given byy*

$$TX = Y \quad \textit{where} \quad Y(t) = X(t) - \int_s^t b(\sigma,X(\sigma))d\sigma.$$

(b *is as before*). *Then*

$$Q_{s,x}T^{-1} = P_{s,x}.$$

Proof. Define $Y(t,w) = X(t,Tw)$ where X is a Brownian motion. We prove that Y is a Brownian motion with respect to $Q_{s,x}$. Clearly Y is progressively measurable because T is $(\mathcal{F}_t - \mathcal{F}_t)$-measurable for every t, i.e. $T^{-1}(\mathcal{F}_t) \subset \mathcal{F}_t$ and X is progressively measurable. Clearly $Y(t,w)$ is continuous $\forall\, w$. We have only to show that $Y(t_2) - Y(t_1)$ is $Q_{s,x}$ -independent of $\mathcal{F}^s_{t_1}$ and has distribution $N(0;(t_2 - t_1)I)$ for each $t_2 > t_1 \geq s$. But we have checked that

$$\exp\left[< \theta, X_t - x > - \frac{1}{2}|\theta|^2(t-s) - \int_s^t < \theta, b > d\sigma\right]$$

is a martingale relative to $Q_{s,x}$. Therefore

$$E^{Q_{s,x}}(\exp < \theta, Y_{t_2} - Y_{t_1} >|\ {}^{s}_{t_1}) = \exp(\tfrac{1}{2}|\theta|^2(t_2 - t_1)),$$

showing that $Y_{t_2} - Y_{t_1}$ is independent of $\mathcal{F}^s_{t_1}$ and has normal distribution $N(0; (t_2 - t_1)I)$. Thus Y is a B rownian motion relative to $Q_{s,x}$. Therefore

$$Q_{s,x}\, T^{-1} = P_{s,x}.$$

21. INTEGRAL EQUATIONS

DEFINITION. A function $b\colon R^d \to R^d$ is said to be locally Lipschitz if given any $x_0 \in R^d$ there exists an open set U_0 containing x_0 such that $b|_{U_0}$ is Lipschitz.

Exercise 1. b is locally Lipschitz iff $b|_K$ is Lips chitz for every compact set K i.e. iff $b|_K$ is Lipschitz for every closed sphere K.

Exercise 2. Every locally Lipschitz function is continuous..

THEOREM. *Let* $b\colon R^d \to R^d$ *be locally Lipschitz and* $X\colon [0,\infty) \to R^d$ *continuous. Then*

(i) *the equation*

$$Y(t) = X(t) + \int_0^t b(Y(s))ds \quad (*)$$

has a continuous solution near 0, *i.e. there exists an* $\varepsilon > 0$ *and a continuous function* $Y\colon [0,\varepsilon] \to R^d$ *such that the above equation is satisfied for all* t *in* $[0,\varepsilon]$.

(ii) *(Uniqueness) If* Y_1, Y_2 *are continuous solutions of the above equation in* $[0,T]$, *then*

$$Y_1 = Y_2 \text{ on } [0,T].$$

Proof. (ii) (Uniqueness) Let $f(t) = |Y_1(t) - Y_2(t)|$. As Y_1, Y_2 are continuous, there exists a $k > 0$ such that $|Y_1(t)|, |Y_2(t)| \le k$ for all t in $[0,T]$. Choose C such that $|b(x) - b(y)| \le C|x-y|$ for $|x|, |y| \le k$ and $C \quad \sup_{0 \le t \le T} f(t)$. Then $f(t) \le C$ and

$f(t) \leq C \int_0^t f(s)ds$ so that $f(t) \leq \frac{(ct)^n}{n!}$ for all $n = 1,2,3,\ldots$,

Thus $Y_1(t) = Y_2(t)$, proving uniqueness.

(i) (Existence) We can very w ell assume that $X(0) = 0$.

Let $a = \inf \{t: |X(t)| \geq 1\}$,

$$M > \sup \{|b(x)|: |x| \leq 2\} , \alpha = \inf \{a, \frac{1}{M}\} ,$$

$C \neq 0$, a Lipschitz constant, so that $|b(x) - b(y)| \leq C |x-y|$ for all $|x|,|y| \leq 2$. Define the iterations $Y_0, Y_1,\ldots$ by

$$Y_0(t) = X(t), Y_{n+1}(t) = X(t) + \int_0^t b(Y_n(s))ds$$

for all $t \geq 0$. By induction, each Y_n i s continuous. By induction again, $|Y_n(t) - X(t)| \leq Mt$ for all n, $0 \leq t \leq \alpha$. Again, by induction $|Y_{n+1}(t) - Y_n(t)| \leq \frac{M}{C} \frac{(Ct)^{n+1}}{(n+1)!}$ for $0 \leq t \leq \alpha$. Thus $Y_n(t)$ converges uniformly on $[0,\alpha]$ to a continuous function $Y(t)$ which is seen to satisfy the integral equation.

REMARK. Let $X: [-\delta,\infty) \to R^d$ be continuous where $\delta > 0$. Then a similar proof guarantees that the equation (*) has a solution in $[-\varepsilon,\varepsilon$ for some $\varepsilon > 0$.

Define $B(X) = \sup \{t:$ (*) has a solution in $[0,t]\}$. The theorem above implies that $0 < B(X) \leq \infty$. $B(X)$ is called the *exploding time*.

REMARK. If b is, in addition, either bounded or globally Lipschitz $B(X) = \infty$ for every continuous $X: [0,\infty) \to \mathbb{R}^d$.

Example. Let $b(y) = y^2$, $X(t) = 0$. The equation

$$Y(t) = x_0 + \int_0^t b(y(s))ds$$

with $x_0 > 0$ has a solution

$$Y(t) = \frac{1}{\frac{1}{x_0} - t}, \ \forall\, t < \frac{1}{x_0};$$

the solution explodes at $t = x_0^{-1}$.

PROPOSITION. *If*

$$B(w) < \infty, \quad \textit{then} \quad \underset{t \uparrow B(w)}{\mathrm{Lt}} |y(t)| = +\infty .$$

Proof. Suppose that $\varliminf_{t \to B(w)} |y| = R < \infty$. Let (t_n) be a sequence increasing to $B(w)$ such that $|y(t_n)| \le R+1$, $\forall$ n. Let

$$\tau_n = \inf\{t \ge t_n : |y(t) - y(-_n)| \ge 1\}.$$

Then

$$1 = |y(\tau_n) - y(t_n)|$$

$$\le w(\tau_n) - w(t_n)| + (\tau_n - t_n) \sup |b(\lambda)| \ .., \quad (1)$$

$$\lambda \in S(y(t_n), 1).$$

Since (t_n) is bounded, we can choose a constant M such that

$$|w(t_n) - w(t)| < \frac{1}{2} \text{ if } |t - t_n| \le M.$$

Then using (1),

$$\tau_n - t_n \ge \inf\{M, (2 \sup |b(\lambda)|)^{-1} \quad \text{where } \lambda \in S(Y(t_n);1)$$

Therefore

$$\tau_n - t_n \geq \inf (M, (2 \sup |b(\lambda)|)^{-1}, \lambda \in S(0;R+2)) = \alpha(\text{say}) \ \forall\ n$$

Chose n such that $\tau_n > B(w) > t_n$. Then y is bounded in $[t_n, B(w)]$ and hence it is bounded in $[0,B(w))$. From the equation

$$y(t) = X(t) + \int_0^t b(y(s))ds$$

one then gets that $\mathop{Lt}_{t \to B(w)} y(t)$ exists. But this is clearly a contradiction since in such a case the solution exists in $[0,B(w)+\varepsilon)$ for a suitable ε, contradicting the definition of $B(w)$. Thus

$$\underline{\lim}_{t \to B(w)} |y(t)| = +\infty$$

and hence

$$\lim_{t \to B(w)} |y(t)| = +\infty.$$

COROLLORY. *If* b *is locally Lipschitz and bounded, then* $B(X) = \infty$ *for all* X *in* $C([0,\infty), R^d)$.

Proof. Left as an exercise.

PROPOSITION. *Let* $b: R^d \to R^d$ *be locally Lipschitz and bounded. Define* $T: C([0,\infty),R^d) \to C([0,\infty),R^d)$ *by* $TX = Y$ *where*

$$Y(t) = X(t) + \int_0^t b(Y(s))ds.$$

Then T *is continuous.*

Proof. Let $X, X^*: [0,\infty) \to R^d$ be continuous, $K > 0$ be given. Let $Y_0, Y_1, ..; Y_0^*, Y_1^*, ...$ be the iterations for X, X^* respectively. Then $|Y_n(t) - X(t)| \leq K \|b\|_\infty$ for $0 \leq t \leq K$, $n = 0,1,2,3,...,$ so that we can find R such that $|Y_n(t)|, |Y_n^*(t)| \leq R$ for $0 \leq t \leq k$,

$n = 0,1,2,\ldots,$ Let $C \geq 1$ be any Lipschitz constant for the function b on $|x| \leq R$. Then

$$|Y_n(t) - Y_n^*(t)| \leq \sup_{0 \leq t \leq K} |X(t)-X^*(t)|.(1+Ct+ \frac{(Ct)^2}{2!} + \\ + .. + \frac{(Ct)^n}{n!}) \quad \text{for } 0 \leq t \leq K, \quad n = 0,1,2,3,\ldots .$$

As b is bounded, Y_n converges uniformly to Y on $[0,K]$. Letting $n \to \infty$, we get

$$\sup_{0 \leq t \leq K} |(TX)t - TX^*)t| \leq e^{cK} \sup_{0 \leq t \leq K} |X(t)-X^*(t)|,\ldots(2)$$

where c depends on $\sup_{0 \leq t \leq K} |X(t)|$, $\sup_{0 \leq t \leq K} |X^*(t)|$. The proof follows by (2).

22. LARGE DEVIATIONS

LET P_ε BE THE Brownian motion starting from zero scaled to Brownian motion corresponding to the operator $\varepsilon\frac{\Delta}{2}$. More precisely, let

$$P_\varepsilon(A) = P\left(\frac{A}{\sqrt{\varepsilon}}\right)$$

where P is the Brownian motion starting at time 0 at the point 0.

<u>INTERPRETATION 1.</u> Let $\{X_t : t \geq 0\}$ be a Brownian motion with $X(0) = x$. Let $Y(t) = X(\varepsilon t)$, $\forall\, t \geq 0$. Then P_ε is the measure induced by the process $Y(t)$. This amounts to stretching the time or scaling time.

<u>INTERPRETATION 2.</u> Let $Y(t,.) = \sqrt{\varepsilon}X(t,.)$. In this case also P_ε is the measure induced by the process $Y(t,.)$. This amounts to 'looking at the process from a distance' or scaling the length.

<u>Exercise.</u> Make the interpretations given above precise.
(Hint: Calculate (i) the probability that $X(\varepsilon t) \in A$, and (ii) the probability that $\sqrt{\varepsilon}\, X(t,) \in A$).

<u>Problem.</u> Let

$$I(w) = \frac{1}{2}\int_0^1 |\dot{w}(t)|^2 dt$$

if $w(0) = 0$, w absolutely continuous on $[0,1]$. Put $I(w) = \infty$ otherwise. We would like to evaluate

$$\int_\Omega e^{\frac{F(w)}{\varepsilon}}\, dP_\varepsilon(w)$$

for small values of ε. Here $F(w): C[0,1] \to \mathbb{R}$ is assumed to be a bounded and continuous function.

<u>THEOREM.</u> *Let* C *be any closed set in* $C[0,1]$ *and let* G *be any open set in* $C[0,1]$. *Then*

$$\limsup_{\varepsilon \to 0} \varepsilon \log P_\varepsilon(C) \leq - \inf_{w \in C} I(w),$$

$$\liminf_{\varepsilon \to 0} \varepsilon \log P_\varepsilon(G) \geq - \inf_{w \in G} I(w).$$

Here $P_\varepsilon(G) = P_\varepsilon(\pi^{-1}G)$ *where* $\Pi: C[0,\infty) \to C[0,1]$ *is the canonical projection.*

<u>*Significance of the theorem.*</u> **1.** If

$$dP_\varepsilon = e^{\frac{-I(w)}{\varepsilon}},$$

then

$$P_\varepsilon(A) = \int_A e^{\frac{-I(w)}{\varepsilon}} dP_\varepsilon$$

is asymptotically equivalent to

$$\exp\left[-\frac{1}{\varepsilon} \inf_{w \in A} I(w)\right].$$

2. If A is any set such that

$$\inf_{w \in A^o} I(w) = \inf_{w \in \bar{A}} I(w),$$

then by the theorem

$$\operatorname*{Lt}_{\varepsilon \to 0} \log P_\varepsilon(A) = \inf_{w \in A} I(w).$$

Proof of the theorem.

LEMMA 1. *Let* $w_0 \in \Omega$ *with* $I(w_0) = \ell < \infty$. *If* $S = S(w_0;\delta)$ *is any sphere of radius* δ *with centre at* w_0 *then* $\underline{\lim}_{\varepsilon \to 0} \varepsilon \log P_\varepsilon(S) \geq -I(w_0)$

Proof.

$$P\,(S) = \int \chi_{S(w_0,\delta)}(w)\, dP_\varepsilon$$

$$= \int \chi_{S(0;\ )}(w-w_0)\, dP_\varepsilon$$

$$= \int \chi_{S(0;\delta)}(\lambda w)\; dP\left(\frac{w}{\sqrt{\varepsilon}}\right), \quad \text{where} \quad \lambda(w) = w - w_0,$$

$$= \int \chi_{S(0;\delta)}(\lambda(\sqrt{\varepsilon}\, w))\; dP(w)$$

$$= \int \chi_{S(0,\delta)}(\sqrt{\varepsilon}\, w) \exp\left[\int_0^1 < \dot{w}_0, dX > - \frac{1}{2}\int_0^1 |\dot{w}_0|^2 d\sigma\right] dP(w$$

$$= \int \chi_{S(0;\delta)}(\sqrt{\varepsilon}\, w) \exp\left[\int_0^1 < \dot{w}_0, dX > - I(w_0)\right] dP(w)$$

$$= \int \chi_{S(0;\delta)}(w) \exp\left[-\frac{1}{\varepsilon}\int_0^1 < \dot{w}_0, dX > - \frac{1}{\varepsilon} I(w_0)\right] dP_\varepsilon(w)$$

$$= e^{\frac{-I(w_0)}{\varepsilon}} P_\varepsilon(S(0;\delta)) \frac{1}{P_\varepsilon(S(0;\delta))} \int_{S(0;\delta)} \exp\left[-\frac{1}{\varepsilon}\int_0^1 < \dot{w}_0, dX >\right] dP_\varepsilon$$

$$\geq e^{\frac{-I(w_0)}{\varepsilon}} P_\varepsilon(S(0;\delta))\; e\left[-\frac{1}{\varepsilon}\,\frac{1}{P_\varepsilon(S(0;\delta))} \int_{S(0;\delta)} \int_0^1 < \dot{w}_0, dX > dP_\varepsilon\right]$$

by Jensen;s inequality,

$$= e^{\frac{-I(w_0)}{\varepsilon}} P_\varepsilon(S(0,\delta))\, e^{0} \quad (\text{use } P_\varepsilon(w)=P_\varepsilon(-w) \text{ if } w \in S(0;\delta))$$

$$= e^{\frac{-I(w_0)}{\varepsilon}} P_\varepsilon(S(0,\delta)).$$

Therefore

$$P_\varepsilon(S(w_0;\delta)) \geq e^{\frac{-I(w_0)}{\varepsilon}} P(S(0; \tfrac{\delta}{\sqrt{\varepsilon}}))$$

or,

$$\varepsilon \log P_\varepsilon(S(w_0;\delta)) \geq -I(w_0) + \varepsilon \log P(S(0;\tfrac{\delta}{\sqrt{\varepsilon}}));$$

let $\varepsilon \to 0$ to get the result. Note that the Lemma is trivially satisfied if $I(w_0) = +\infty$.

Proof of Part 2 of the theorem.

Let G be open, $w_0 \in G$; then there exists $\delta > 0$ with $S(w_0,\delta) \subset G$. By Lemma 1

$$\underline{\lim}_{\varepsilon \to 0} \varepsilon \log P_\varepsilon(G) \geq \underline{\lim}_{\varepsilon \to 0} \varepsilon \log P_\varepsilon(S(w_0;\delta)) \geq -I(w_0).$$

Since w_0 is arbitrary, we get

$$\underline{\lim}\ \varepsilon \log P_\varepsilon(G) \geq -\inf\{I(w_0): w_0 \in G\}.$$

For Part 1 we need some more preliminaries.

LEMMA 2. *Let* $(w_n) \in C[0,1]$ *be such that* $w_n \to w$ *uniformly on* $[0,1], I(w_n) \leq \alpha < \infty$. *Then* $I(w) < \alpha$, *i.e.* I *is lower semi-continuous.*

Proof. Step 1. w is absolutely continuous. Let $\{(x_i', x_i'')\}_{i=1}^{n}$ be a collection of mutually disjoint intervals in $[0,1]$. Then

$$\sum_{i=1}^{n} |w_m(x_i') - w_m(x_i'')| \le \sum_{i=1}^{n} |x_i'' - x_i'|^{1/2} [\int_{x_i'}^{x_i''} |w_m|^2]^{1/2}$$

(by Hölder's inequality)

$$\le (\sum_{i=1}^{n} \int_{x_i'}^{x_i''} |w_m|^2)^{1/2} (\sum_{i=1}^{n} |x_i'' - x_i'|)^{1/2} \quad \text{(again by Hölder)}$$

$$\le \sqrt{(2\alpha)} (\sum |x_i'' - x_i'|)^{1/2}.$$

Letting $m \to \infty$ we get the result.

<u>Step 2.</u> Observe that $w_m(0) = 0$ S$_0$w(0) = 0. Therefore

$$\left|\frac{w_n(x+h) - w_n(x)}{h}\right|^2 = \left|\frac{1}{h} \int_x^{x+h} w_n dt\right|^2 \le \frac{1}{h^2} (\int_x^{x+h} |w_n| dt)^2$$

$$\le \frac{1}{h} \int_x^{x+h} |w_n|^2 dt.$$

Hence

$$\int_0^{1-h} \left|\frac{w_n(x+h) - w_n(x)}{h}\right|^2 dx \le \frac{1}{h} \int_0^{1-h} dx \int_0^h |(\dot{w}_n(x+t))|^2 dt$$

$$\le \frac{1}{h} \int_0^h dt \int_0^{1-h} |\dot{w}_n(x+t)|^2 dx$$

$$\le \frac{1}{2}\, 2 \int_0^h dt = 2\alpha$$

letting $n \to \infty$, we get

$$\int_0^{1-h} \left|\frac{w(x+h) - w(x)}{h}\right|^2 dx \le 2\alpha.$$

Let $h \to 0$ to get $I(w) \le \alpha$, completing the proof.

LEMMA 3. *Let* C *be closed and put* $C^{\delta} = \bigcup_{w \in C} S(w;\delta)$; *then*

$$\lim_{\delta \to 0} \left(\inf_{w \in C^{\delta}} I(w) \right) = \inf_{w \in C} I(w).$$

Proof. If $\delta_1 < \delta_2$, then $C^{\delta_1} \subset C^{\delta_2}$ so that $\inf_{w \in C^{\delta}} I(w)$ is decreasing. As $C^{\delta} \supset C$ for each δ,

$$\lim_{\delta \to 0} \left(\inf_{w \in C^{\delta}} I(w)\right) \leq \inf_{w \in C} I(w)$$

Let $\ell = \lim_{\delta \to 0} \left(\inf_{w \in C^{\delta}} I(w)\right)$. Then there exists $w_{\delta} \in C^{\delta}$ such that $I(w_{\delta}) \to \ell$, and therefore $(I(w_{\delta}))$ is a bounded set bounded by α (say).

CLAIM.

$$|w_{\delta}(t_1) - w_{\delta}(t_2)| = \left| \int_{t_1}^{t_2} \dot{w}_{\delta} dt \right| \leq \sqrt{(|t_1 - t_2|)} \left(\int |\dot{w}_{\delta}|^2\right)^{1/2}$$

$$\leq \sqrt{(2\alpha|t_1 - t_2|)}.$$

The family (w_{δ}) is therefore equicontinuous which, in view of the fact that $w_{\delta}(0) = 0$, implies that it is uniformly bounded and the claim follows from Ascoli's theorem. Hence every subfamily of (w_{δ}) is equicontinuous. By Ascoli's theorem there exists a sequence $\delta_n \to 0$ such that $w_{\delta_n} \to w$ uniformly on $[0,1]$. It is clear that $w \in C$. By lower semicontinuity of $I(w)$,

$$\lim_{\delta \to 0} \inf_{w \in C^{\delta}} I(w) \geq \inf_{w \in C}$$

completing the proof.

Proof of Part 1 of the theorem. Let X be continuous in [0.1]. For each n let X_n be a piecewise linear version of X bas ed on n equal intervals, i.e. X_n is a polygonal function joining the points $(0,X(0)), (1/n,X(1/n)),\ldots,(1,X(1))$.

$$P_\varepsilon(\|X_n - X\| \geq \delta) \ , \ (\|\cdot\| = \|\cdot\|_\infty)$$

$$\leq P\Big(\bigcup_{n} \sup_{i \leq j \leq n} \cdot \sup_{\frac{j-1}{n} \leq t \leq \frac{j}{n}} |X_r(t) - X_r(\tfrac{j-1}{n})| \geq \frac{\delta}{2\sqrt{d}}\Big),$$

where $X = (X_1,\ldots,X_d)$.

$\leq nd\, P_\varepsilon(\sup_{0 \leq t \leq 1/n} |X(t) - X(0)| \geq \frac{\delta}{2\sqrt{d}}$ (Markov property; here X is one-dimensional).

$\leq nd\, P_\varepsilon(\sup_{0 \leq t \leq 1/n} |X_t| \geq \frac{\delta}{2\sqrt{d}})$ (since $X(0) = 0$)

$\leq 2nd\, P_\varepsilon(\sup_{0 \leq t \leq 1/n} X_t \geq \frac{\delta}{2\sqrt{d}})$

$= 2dn\, P(\sup_{0 \leq t \leq 1/n} X_t \geq \frac{\delta}{2\sqrt{\varepsilon}\, d})$

$= 4dn\, P(X(1/n) \geq \frac{\delta}{2\sqrt{\varepsilon}\, d})$ (by the reflection principle)

$$= 4dn \int_{\delta\sqrt{n}/2\sqrt{\varepsilon d}}^{\infty} \frac{1}{\sqrt{2\pi/n}}\, e^{-ny^2/2}\, dy$$

$$= 4dn \int_{\delta\sqrt{n}/2\sqrt{\varepsilon d}}^{\infty} \frac{1}{\sqrt{2\pi}}\, e^{-x^2/2}\, dx$$

Now, for every $a > 0$,

$$a \int_a^\infty e^{-x^2/2}\, dx \leq \int_a^\infty x\, e^{-x^2/2} dx = e^{-a^2/2}.$$

Thus

$$P_\varepsilon(\|X_n - X\| \geq \delta) \leq \frac{4dn}{\sqrt{2\pi}} \frac{e^{-n\delta^2/(8\varepsilon d)}}{\delta\sqrt{n}/2\sqrt{\varepsilon d}} = C_1(n)\frac{\sqrt{\varepsilon}}{\sqrt{\delta}} e^{-n\delta^2/(8\varepsilon d)},$$

where C_1 depends only on n. We have now

$$P_\varepsilon(X_n \in C^\delta) \leq P_\varepsilon(I(X_n) \geq \ell_\delta) \quad \text{where } \ell_\delta = \inf\{I(w) w \in C^\delta\}.$$

$$= P\left(\frac{1}{2}\sum_{j=0}^{n-1} n\left|X\left(\frac{j+1}{n}\right) - X\left(\frac{j}{n}\right)\right|^2 \geq \ell_\delta\right)$$

$$= P\left(Y_1^2 + Y_2^2 + \ldots + Y_{nd}^2 \geq \frac{2\ell_\delta}{\varepsilon}\right),$$

where $Y_1 = \sqrt{n}\,(X_1(1/n) - X_1(0))$ etc. are *independent* normal random variables with mean 0 and variance 1. Therefore,

$$P\left(Y_1^2 + \ldots + Y_{nd}^2 \geq \frac{2\ell_\delta}{\varepsilon}\right)$$

$$= \int_{y_1^2+\ldots+y_{nd}^2 \quad \frac{2\ell_\delta}{\varepsilon}} e^{-(y_1^2+\ldots+y_{nd}^2)^{1/2}} dy_1 \ldots dy_{nd}.$$

$$= C(n)\int_{\sqrt{(2\ell_\delta/\varepsilon)}}^{\infty} e^{-r^2/2}\, r^{nd-1}\, dr,$$

using polar coordinates, i.e.

$$P\left(Y_1^2 + Y_2^2 + \ldots + Y_{nd}^2 \geq \frac{2\ell_\delta}{\varepsilon}\right) = C'(n)\int_{(\ell_\delta/\varepsilon)}^{\infty} e^{-s}\, s^{\frac{nd}{2}-1}\, ds$$

(change the variable from r to $s = \frac{r^2}{2}$). An integration by parts gives

$$\int_\alpha^\infty e^{-s} s^k ds = e^{-\alpha}\left(\alpha^k + \frac{k!}{(k-1)!}\alpha^{k-2} + \ldots\right).$$

Using this estimate (for n even) we get

$$P\left((Y_1^2 + \ldots + Y_{nd}^2) \geq \frac{2\ell_\delta}{\varepsilon}\right) \leq C_2(n) e^{-\ell_\delta/\varepsilon}\left(\frac{\ell_\delta}{\varepsilon}\right)^{\frac{nd}{2}-1},$$

where C_2 depends only on n. Thus,

$$P_\varepsilon(C) \le P_\varepsilon(\| X_n - X \| \ge \delta) + P_\varepsilon(X_n \notin C^\delta)$$

$$\le C_1(n)\sqrt{(\tfrac{\varepsilon}{\delta})}e^{-n\delta^2/(8\varepsilon d)} + C_2(n)e^{-\ell\delta/\varepsilon}(\tfrac{\ell_\delta}{\varepsilon})^{\frac{nd}{2}-1}$$

$$\le 2 \max [C_1(n)\sqrt{(\tfrac{\varepsilon}{\delta})}e^{-n\delta^2/(8\varepsilon d)}, C_2(n)e^{-\ell\delta/\varepsilon}(\tfrac{\ell_\delta}{\varepsilon})^{\frac{nd}{2}-1}$$

$$\varepsilon \log P_\varepsilon(C) \le \varepsilon \log 2 + \varepsilon \max[\log (C_1(n)\sqrt{(\tfrac{\varepsilon}{\delta})}e^{-n\delta^2/(8\varepsilon d)}$$

$$\log C_2(n)\, e^{-\ell_\delta/\varepsilon}(\tfrac{\ell_\delta}{\varepsilon})^{\frac{nd}{2}-1}]$$

Let $\varepsilon \to 0$ to get

$$\overline{\lim}\, \varepsilon \log P_\varepsilon(C) \le \max \{\frac{-n\delta^2}{8d}, \frac{-\ell_\delta}{1}\}.$$

Fix δ and let $n \to \infty$ through even values to get

$$\overline{\lim}\, \varepsilon \log P_\varepsilon(C) \le -\ell_\delta.$$

Now let $\delta \to 0$ and use the previous lemma to get

$$\overline{\lim_{\varepsilon \to 0}}\, \varepsilon \log P_\varepsilon(C) \le -\inf_{w \in C} I(w).$$

PROPOSITION. *Let* ℓ *be finite; then* $\{w: I(w) \le \ell\}$ *is compact in* Ω.

Proof. Let (w_n) be any sequence, $I(w_n) \le \ell$. Then

$$|w_n(t_1) - w_n(t_2)| \le \sqrt{(\ell|t_1 - t_2|)}$$

and since $w_n(0) = 0$, we conclude that $\{w_n\}$ is equicontinuous and uniformly bounded.

ASSUMPTIONS. Let Ω be any separable metric space, $\mathcal{F}$ = Borel σ-field on Ω. For every $\varepsilon > 0$ let P_ε be a probab ility measure. Let $I: \Omega \to [0,\infty]$ be any function such that

(i) I is lower semi-continuous.

(ii) $\forall$ finite ℓ, $\{w: I(w) \leq \ell\}$ is compact.

(iii) For every closed set C in Ω,

$$\lim_{\varepsilon \to 0} \sup \varepsilon \log P_\varepsilon(C) \leq -\inf_{w \in C} I(w).$$

(iv) For every open set G in Ω

$$\lim_{\varepsilon \to 0} \inf \varepsilon \log P_\varepsilon(G) \geq -\inf_{w \in G} I(w).$$

REMARK. Let $\Omega = C[0,1]$, P_ε the Brownian measure corresponding to the scaling ε. If $I(w) = \frac{1}{2}\int_0^1 |w|^2 dt$ if $w(0) = 0$ and ∞ otherwise, then all the above assumptions are satisfied.

THEOREM. *Let* $F: \Omega \to \mathbb{R}$ *be bounded and continuous. Under the above assumptions the following results hold.*

(i) For every closed set C *in* Ω.

$$\lim_{\varepsilon \to 0} \sup \varepsilon \log \int_C \exp \frac{F(w)}{\varepsilon} dP_\varepsilon \leq \sup_{w \in C}(F(w)-I(w)).$$

(ii) For every open set G *in* Ω

$$\lim_{\varepsilon \to 0} \inf \varepsilon \log \int_G \exp \frac{F(w)}{\varepsilon} dP_\varepsilon \geq \sup_{w \in G} (F(w)-I(w)).$$

In particular, if $G = \Omega = C$, *then*

$$\lim_{\varepsilon \to 0} \varepsilon \log \int_\Omega \exp \frac{F(w)}{\varepsilon} dP_\varepsilon = \sup_{w \in \Omega} (F(w)-I(w)).$$

Proof. Let G be open, $w_0 \in G$. Let $\delta \to 0$ be given. Then there exists a neighbourhood N of w_0, $F(w) \geq F(w_0) - \delta$, $\forall$ w in N. Therefore

$$\int_G \exp \frac{F(w)}{\varepsilon}\, dP_\varepsilon \geq \int_N \exp \frac{F(w)}{\varepsilon}\, dP_\varepsilon \geq e^{\frac{F(w_0)-\delta}{\varepsilon}} P_\varepsilon(N).$$

Therefore

$$\varepsilon \log \int_G \exp \frac{F(w)}{\varepsilon}\, dP_\varepsilon \geq F(w_0) - \delta + \varepsilon \log P_\varepsilon(N).$$

Thus

$$\underline{\lim} \quad \log \int_G \exp \frac{F(w)}{\varepsilon}\, dP_\varepsilon \geq F(w_0) - \delta + \underline{\lim}\, \varepsilon \log P_\varepsilon(N).$$

$$\geq F(w_0) - \delta - \inf_{w \in N} I(w) \geq F(w_0) - I(w_0) - \delta.$$

Since δ and w_0 are arbitrary ($w_0 \in G$) we get

$$\underline{\lim}\, \varepsilon \log \int_G \exp \frac{F(w)}{\varepsilon}\, dP_\varepsilon \geq \sup_{w \in G} (F(w) - I(w)).$$

This proves Part (ii) of the theorem.

<u>Proof of Part (i)</u>. <u>Step 1</u>. Let C be compact; $L = \sup_{w \in G} (F(w)-I(w))$.

If $L = -\infty$ it follows easily that

$$\lim_{\varepsilon \to 0} \sup \varepsilon \log \int_C e^{F/\varepsilon} dP_\varepsilon \leq -\infty.$$

(Use the fact that F is bounded). Thus without any loss, we may assume L to be finite. Let $w_0 \in C$; then there exists a neighbourhoo[d] N of w_0 such that $F(w) \leq F(w_0) + \delta$ and b y lower semi-continuity of I,

$$I(w) \geq I(w_0) - \delta, \ \forall\, w \in N(w_0).$$

By regularity, there exists an open set G_{w_0} containing w_0 such that

$G_{w_0} \quad \overline{G}_{w_0} \quad N(w_0)$. Therefore

$$\int_{\overline{G}_{w_0}} \exp \frac{F(w)}{\varepsilon}\, dP_\varepsilon \quad \exp\left(\frac{F(w_0)+\delta}{\varepsilon}\right) P_\varepsilon(\overline{G}_{w_0}).$$

Therefore

$$\limsup_{\varepsilon \to 0} \varepsilon \log \int_{G_{w_0}} \exp \frac{F(w)}{\varepsilon}\, dP_\varepsilon \le F(w_0) + \delta + \varepsilon \overline{\lim_{\varepsilon \to 0}}\, P_\varepsilon(\overline{G}_{w_0})$$

$$\le F(w_0) + \delta - \inf_{w \in \overline{G}_{w_0}} I(w)$$

$$\le F(w_0) + \delta - I(w_0) + \delta$$

$$\le L + 2\delta .$$

Let $K_\ell = \{w: I(w) \le \ell\}$. By assumption, K_ℓ is compact. Therefore, for each $\delta > 0$, there exists an open set G_δ containing $K_\ell \cap C$ such that

$$\limsup_{\varepsilon \to 0} \varepsilon \log \int_{G_\delta} e^{\frac{F(w)}{\varepsilon}}\, dP_\varepsilon \le L + 2\delta.$$

Therefore

$$\limsup_{\varepsilon \to 0} \varepsilon \log \int_{G_\delta \cap C} e^{\frac{F(w)}{\varepsilon}}\, dP_\varepsilon \le L + 2\delta,$$

$$\int_{G_\delta^c \cap C} e^{\frac{F(w)}{\varepsilon}}\, dP_\varepsilon \le e^{M/\varepsilon}\, P\,(C \cap G_\delta^c).$$

Therefore

$$\limsup_{\varepsilon \to 0} \varepsilon \log \int_{G_\delta^c \cap C} e^{\frac{F(w)}{\varepsilon}}\, dP_\varepsilon \le M + \limsup_{\varepsilon \to 0} \varepsilon \log P_\varepsilon(G_\delta^c \cap C)$$

$$\le M - \inf_{w \in C \cap G_\delta^c} I(w).$$

Now

$$G_\delta^c \subset K_\ell^c \cup C^c.$$

Therefore

$$C \cap G_\delta^c \subset C \cap K_\ell^c$$

if $w \in C \cap G_\delta^c$, $w \notin K_\ell$. Therefore $I(w) > \ell$. Thus

$$\lim_{\varepsilon \to 0} \sup \varepsilon \log \int_{G_\delta^c \cap C} e^{F(w)/\varepsilon} dP_\varepsilon \leq M-\ell \leq L \leq L + 2\delta.$$

This proves that

$$\lim_{\varepsilon \to 0} \sup \varepsilon \log \int_C \exp \frac{F(w)}{\varepsilon} dP_\varepsilon \leq L + 2\delta.$$

Since C is compact there exists a finite number of points $w_1,\ldots,w_n$ in C such that

$$C \subset \bigcup_{i=1}^{n} G_{w_i}$$

Therefore

$$\overline{\lim}\, \varepsilon \log \int_C \exp \frac{F(w)}{\varepsilon} dP_\varepsilon \leq \overline{\lim}\, \varepsilon \log \int_{\bigcup_{i=1}^{n} G_{w_i}} e^{F(w)/\varepsilon} dP_\varepsilon$$

$$\leq \overline{\lim}\, (\varepsilon \log n \max_{1 \leq i \leq n} \int_{G_{w_i}} \exp \frac{F(w)}{\varepsilon} dP_\varepsilon)$$

$$\leq L + 2\delta.$$

Since δ is arbitrary,

$$\overline{\lim}\, \varepsilon \log \int_C \exp \frac{F(w)}{\varepsilon} dP_\varepsilon \leq \sup_{w \in C} (F(w)-I(w)).$$

The above proof shows that given a compact set C, and $\delta > 0$ there exists an open set G containing C such that

$$\overline{\lim}\ \varepsilon \log \int_G \exp \frac{F(w)}{\varepsilon}\, dP_\varepsilon \le L + 2\delta.$$

Step 2. Let C be any arbitrary closed set in Ω. Let

$$L = \sup_{w \in C} (F(w) - I(w)).$$

Since F is bounded there exists an M such that $|F(w)| \le M$ for all w. Choose ℓ so large that $M - \ell \le L$. Since δ is arbitrary

$$\lim_{\varepsilon \to 0} \sup \varepsilon \log \int_C \exp \frac{F(w)}{\varepsilon}\, dP_\varepsilon \le \sup_{w \in C} (F(w)-I(w))$$

We now prove the above theorem when P_ε is replaced by Q_x^ε. Let P_x^ε be the Brownian motion starting at time $t = 0$ at the space point x corresponding to the scaling ε. Precisely stated, if

$$\tau_\varepsilon : C([0,\infty); \mathbb{R}^d) \to C([0,\infty); \mathbb{R}^d)$$

is the map given by $(\tau_\varepsilon w)(t) = w(\varepsilon t)$, then $P_x^\varepsilon \underset{\text{def}}{=} P_x \tau_\varepsilon^{-1}$. Note

$T_1''\tau_\varepsilon = T_\varepsilon$ and T_ε is given by

$$T_\varepsilon w = y \quad \text{where} \quad y(t) = w(\varepsilon t) + \int_0^t b(y(s))ds.$$

Hence

$$P_x T_\varepsilon^{-1} = P_x(T_1,\tau_\varepsilon)^{-1} = P_x \tau_\varepsilon^{-1} T_1^{-1} = P_x^\varepsilon\, T_1^{-1};$$

either of these probability measures is denoted by Q_x^ε.

THEOREM. *Let* $b\colon \mathbb{R}^d \to \mathbb{R}^d$ *be bounded measurable and locally Lipschitz. Define*

$$I(w): \frac{1}{2} \int_0^1 |\dot{X}(t) - b(X(t))|^2 dt$$

If $w \in C([0,\infty); \mathbb{R}^d)$, $w(0) = x$ *and* x *absolutely continuous. Put* $I(w) = \infty$ *otherwise. If* C *is closed in* $C([0,1]; \mathbb{R}^d)$, *then*

$$\overline{\lim_{\varepsilon \to 0}} \; \varepsilon \log Q_x^\varepsilon(C) \leq - \inf_{w \in C} I(w).$$

If G *is open in* $C([0,1]; \mathbb{R}^d)$, *then*

$$\underline{\lim}_{\varepsilon \to 0} \; \varepsilon \log Q_x^\varepsilon(G) \geq - \inf_{w \in G} I(w).$$

As usual $Q_x^\varepsilon(C) = Q_x^\varepsilon \pi^{-1}(C)$ *where* $\pi: C([0,\infty); \mathbb{R}^d) \to C([0,1]; \mathbb{R}^d)$ *is the canonical projection.*

REMARK. If $b = 0$ we have the previous case.

Proof. Let T be the map $x(.) \to y(.)$ where

$$y(t) = x(t) + \int_0^t b(y(s))ds.$$

Then

$$Q_x^\varepsilon = P_x^\varepsilon (T^{-1}).$$

If C is closed

$$Q_x^\varepsilon(C) = P_x^\varepsilon(T^{-1}C).$$

The map T is continuous. Therefore $T^{-1}(C)$ is closed. Thus

$$\limsup_{\varepsilon \to 0} \varepsilon \log Q_x(C) = \limsup_{\varepsilon \to 0} \varepsilon \log P_x^\varepsilon(T^{-1}C)$$

(*) $$\leq -\inf_{w \in T^{-1}(C)} \frac{1}{2} \int_0^1 |\dot{X}|^2 dt \qquad \text{(see Exercise 1 below)}$$

$$= -\inf_{w \in C} \frac{1}{2} \int_0^1 |\dot{T^{-1}w}|^2 dt.$$

Now

$$y(.) \overset{T^{-1}}{\to} y(t) - \int_0^t b(y(s))ds.$$

Therefore

$$(T^{-1}y) = y - b(y(s)).$$

Therefore

$$\limsup_{\varepsilon \to 0} \varepsilon \log Q_x^{\varepsilon}(C) \leq -\inf_{w \in C} I(w).$$

The proof when G is open is similar.

<u>Exercise 1</u>. Replace P_ε by P_x^{ε} and I by I_x where

$$I_x(w) = \frac{1}{2} \int_0^1 |w|^2, \; w(0) = x, \; w \text{ absolutely continuous},$$

$$= \infty \quad \text{otherwise}.$$

Check that (*) holds, i.e.

$$\limsup_{\varepsilon \to 0} \varepsilon \log P_x^{\varepsilon}(C) \leq -\inf_{w \in C} I_x(w), \quad \text{if } C \text{ is closed},$$

and

$$\liminf_{\varepsilon \to 0} \varepsilon \log P_x^{\varepsilon}(G) \geq -\inf_{w \in G} I_x(w).$$

Let G be a bounded open set in $\mathbb{R}^n$, with a s mooth boundary $\Gamma = \partial G$. Let $b: \mathbb{R}^d \to \mathbb{R}^d$ be a smooth C^∞ function such that

(i) $\langle b(x), n(x) \rangle > 0$, $\forall x \in \partial G$ where $n(x)$ is the unit inward normal.

(ii) there exists a point $x_0 \in G$ with $b(x_0) = 0$ and $|b(x)| > 0$, $\forall x$ in $G - \{x_0\}$.

(iii) for any x in G the solution

$$\xi(t) = x + \int_0^t b(\xi(s))ds,$$

of the vector field starting from x converges to x_0 as $t \to +\infty$.

REMARK. (a) (iii) is uaually interpreted by saying that "x_0 is stable".

(b) By (i) and (ii) every solution of (iii) takes all its values in G and ultimately stays close to x_0.

Let $\varepsilon > 0$ be given; $f: \partial G \to \mathbb{R}$ be any continuous bounded function. Consider the system

$$L_\varepsilon u_\varepsilon = \frac{1}{2} \Delta u_\varepsilon + b(x).\Delta u_\varepsilon = 0 \quad \text{in } G$$

$$u_\varepsilon = f \quad \text{on } \partial G.$$

We want to study $\lim_{\varepsilon \to 0} u_\varepsilon(x)$. Define

$$I_0^T(X(t)) = \frac{1}{2} \int_0^T |X(t) - b(X(t))|^2 dt; X:[0,T] \to \mathbb{R}^d$$

whenever X is absolutely continuous, $= \infty$ otherwise.

REMARK. Any solution of (iii) is called an integral curve. For any curve X on $[0,T]$, I_0^T gives a measure of the deviation of X from being an integral curve. Let

$$V_T(x,y) = \inf \{I_0^T(X): X(0) = x; X(T) = y\}$$

and

$$V(x,y) = \inf \{V_T(x,y): T > 0\}.$$

V has the following properties.

(i) $V(x,y) \leq V(x,z) + V(z,y) \ \forall\ x,y,z.$

(ii) Given any x, $\exists\ \delta \to 0$ and $C > 0$ such that for all y with $|x - y| \leq \delta$.

$$V(x,y) \leq C|x-y|$$

<u>Proof</u>. Let $X(t) = \dfrac{t(y-x)}{|y-x|} + x.$

Put

$$T = |y - x|,\ X(0) = x,\ X(T) = y,$$

$$I_0^T(X(t)) = \frac{1}{2} \int_0^T \left|\frac{y-x}{T} - b(X + \frac{s}{T}(y-x))\right|^2 ds.$$

Then

$$I_0^T \leq \frac{1}{2} \int_0^T 2\left(\frac{|y-x|^2}{T^2} + \|b\|_\infty^2\right) ds,$$

where

$$\|b\|_\infty = \sup_{|\lambda - x| \leq |y-x|} b(\lambda),$$

or,

$$I_0^T \leq (1 + \|b\|_\infty^2)\,|y - x|.$$

As a consequence of (ii) we conclude that

$$V(x,y) \leq (1 + \sup_{|\lambda-x| \leq |y-x|} |b(\lambda)|^2)\,|y-x|,$$

i.e. V is locally Lipschitz.

The answer to the problem raised is given by the following.

<u>THEOREM</u>. $$\lim_{\varepsilon \to 0} u_\varepsilon(x) = f(y_0)$$

where y_0 *is assumed to be such that* $y_0 \in \partial G$ *and*

$$V(x_0,y_0) < V(x,y),\ \forall\ y \in \partial G,\ y \neq y_0.$$

We first proceed to get an equivalent statement of the theorem. Let P_x^ε be the Brownian measure corresponding to the starting point x, and corresponding to the scaling ε. Then there exis ts a probabil[illegible] measure Q_x^ε such that

$$\left.\frac{dQ_x^\varepsilon}{dP_x^\varepsilon}\right|\mathcal{F}_t = Z(t)$$

where

$$Z(t,.) = \exp \int_0^t < b^*(X(s)),dX(s) > -\frac{1}{2} \int_0^t b^*(X(s))\ ds;$$

b^* is any bounded smooth function such that $b^* = b$ on G. Further we have the integral representation

$$u_\varepsilon(x) = \int_{\partial G} f(X(\tau))\ dQ_x^\varepsilon$$

where τ is the exit time of G, i.e.

$$\tau(w) = \inf\ \{t: w(t) \notin G\}.$$

$$|u_\varepsilon(x) - f(y_0)| = |\int_{\partial G} (f(X(\tau)) - f(y_0))\ dQ_x^\varepsilon|$$

$$\leq |\int_{N \cap \partial G} (f(X(\tau)) - f(Y_0))dQ_x^\varepsilon| +$$

$$+ |\int_{N^c \cap \partial G} (f(X(\tau)) - f(Y_0))\ dQ_x^\varepsilon|$$

(N is any neighbourhood of y_0).

$$\leq Q_x^\varepsilon(X(\tau) \in N \cap \partial G) \sup_{\lambda \in N \cap \partial G} |f(\lambda) - f(y_0)| +$$

$$+ 2 \|f\|_\infty Q_x^\varepsilon(X(\tau) \in N^c \cap \partial G).$$

Since f is continuous, to prove the theorem it is sufficient to prove the

THEOREM. $\lim_{\varepsilon \to 0} Q_x(X(\tau) \in N^c \cap \partial G) = 0$

for every neighbourhood N *of* y_0.

Let N be any neighbourhood of y_0. Let

$$V = V(x_0,y_0), \quad V' = \inf_{y \in N^c \cap \partial G} V(x,y).$$

By definition of y_0 and the fact that $N^c \cap \partial G$ is compact, we conclude that $V' > V$. Choose $\eta = \eta(N) > 0$ such that $V' = V + \eta$. For any $\delta > 0$ let $D = S(x_0;\delta) = \{y: |y - x_0| < \delta\}$, $\partial D = \{y: |y - x_0| = \delta\}$.

CLAIM. We can choose a δ_2 such that

(i) $V(x,y) \geq V + \frac{3\eta}{4}$, $\forall x \in \partial D_2$, $y \in N^c$ ∂G.

(ii) $V(x,y_0) \leq V + \frac{\eta}{4}$, $\forall x \in \partial D_2$.

Proof. (i) $V(x_0,y) \geq V + \eta$, $\forall y \in N^c$ ∂G. Therefore

$$V + \eta \leq V(x_0,y) \leq V(x_0,x) + V(x,y)$$

$$\leq C |x - x_0| + V(x,y)..$$

Choose C such that $C |x - x_0| \leq \frac{\eta}{4}$. Thus

$$V + \frac{3\eta}{4} \leq V(x,y) \quad \text{if} \quad C\,|x-x_0| \leq \frac{\eta}{4}\,, \ \forall\ y \in N^c \quad \partial G.$$

C depends only on x_0. This proves (i).

(ii) $|V(x_0,y_0) - V(x,y_0)| \leq V(x_0,x) \leq C|x_0-x| \leq \frac{\eta}{4}$ if x is close to x_0. Thus

$$V(x,y_0) \leq V(x_0,y_0) + \frac{\eta}{4} = V + \frac{\eta}{4}$$

if x is close to x_0. This can be achieved by choosing δ_2 very small.

CLAIM (iii). We can choose $\delta_1 < \delta_2$ such that for points x_1,x_2 in ∂D_1 there is a path $X(.)$ joining x_1,x_2 with $X(.) \in D_2 - D_1$, i.e. it never penetrates D_1; and

$$I(X) \leq \frac{\eta}{8}\,.$$

Proof. Let $C = \sup\{|b(\lambda)|^2 : |\lambda-x_0| \leq \delta_2\}$. Choose $X(.)$ to be any path on $[0,T]$, taking values in D_2 with $X(0) = x_1$; $X(T) = x_2$ and such that $|\dot{X}| = 1$ (i.e. the path has unit speed). Then

$$I_0^T(X) \leq \int_0^T (|\dot{X}|^2 + C)dt \leq CT + \int_0^T |\dot{X}|dt$$

$$= (c+1)T = (C+1)\,|x_2-x_1|.$$

Choose δ_1 small such that $(C+1)\,|x_2-x_1| \leq \frac{\eta}{8}$.

Let $\Omega\delta_1 = \{w : w(t) \in \overline{G} - D_1,\ \forall\ t \geq 0\}$, i.e. $\Omega\delta_1$ consists of all trajectories in $\overline{G}$ that avoid D_1.

Claim (iv). $\inf_{T>0}\ \inf_{X\in\Omega_{\delta_1},\,X(0)\in\partial D_2,\ X(T)\in N^c\cap\partial G} I_0^T(X(.)) \geq V + \frac{3\eta}{4}$

Proof. Follows from Claim (i) and (ii).

Claim (v). $\inf_{T>0}\ \inf_{X\in\Omega_{\delta_1},\,X(0)\in\partial D_2,\ X(T)=y_0} I_0^T(X(.)) \leq V + \frac{3\eta}{8}$.

Proof. By (ii) $V(x,y_0) \leq V + \frac{\eta}{4}$ $\forall\, x\in\partial D_2$, i.e.

$$\inf_{T>0}\ \inf_{X(0)=x,\,X(T)=y_0} I_0^T(X(.)) \leq V + \frac{\eta}{4}.$$

Let $\varepsilon > 0$ be arbitrary. Choose T and $X(.)$ such that $I_0^T(X) \leq V + \frac{\eta}{4} + \varepsilon$ with $X(0) = x$, $X(T) = y_0$, $X(.)\in\overline{G}$. If $X\in\Omega_{\delta_1}$ define $Y = X$. If $X\notin\Omega_{\delta_1}$ define Y as follows:

Let t_1 be the first time that X enters D_1 and t_2 the last time that it gets out of D_1. Then $0 < t_1 \leq t_2 < T$. Let X^* be a path on $[0,s]$ such that (by Claim (iii)) $I_0^s(X^*) \leq \frac{\eta}{8}$ with $X^*(0) = X(t_1)$ and $X^*(s) = X(t_2)$. Define Y on $[0,T-(t_2-t_1)+ s]$ $[T-(t_2-t_1)+ s,\infty)$ by

$$Y(t) = X(t) \text{ on } [0,t_1] = X^*(t-t_1) \text{ on } [t_1,t_1+s]$$

$$= X(t-t_1-s+t_2), \text{ on } [t_1+s,T-(t_2-t_1)+s]$$

$$= X(t_2), \text{ for } t \geq T-(t_2-t_1)+ s.$$

Then

$$I_0^{T-t_2+t_1+s} = \frac{1}{2}\int_0^{t_1} |X-b(X(s))|^2 ds + \frac{1}{2}\int_0^{s} |X^*-b(X^*(s))|^2 ds$$

$$+ \frac{1}{2}\int_{t_2}^{T} |X(s) - b(X(s))|^2 ds$$

$$\leq V + \frac{\eta}{4} + \epsilon + \frac{\eta}{8}$$

by choice of X and X^*. As $Y \in \Omega_{\delta_1}$, we have shown that

$$\inf_{T>0}\ \inf_{\substack{X \in \Omega_{\delta_1}, X(0) \in \partial D_1 \\ X(T) = y_0}} I_0^T(X(.)) \leq V + \frac{3\eta}{8} + \varepsilon .$$

Since ε is arbitrary we have proved (v).

<u>LEMMA</u>. I_0^T *is lower semi-continuous for every finite* T.

<u>Proof</u>. This is left as an exercise as it involves a repetiti on of an argument used earlier.

<u>LEMMA</u>. *Let* $X_n \in \Omega_{\delta_1}$. *If* $T_n \to \infty$ *then* $I_0^{T_n}(X_n) \to \infty$.

This result says that we cannot have a trajectory which starts outside of a deleted ball for which I remains finite for arbitrary long lengths of time.

<u>Proof</u>. Assume the contrary. Then there exists a constant M such that $I_0^{T_n}(X_n) \leq M$, $\forall$ n. Let $T < \infty$, so that $M_T = \sup_n I_0^T(X_n) < \infty$.

Define $X_n^T = X_n|_{[0,T]}$.

Claim. $\{X_n^T\}_{n=1}$ is an equicontinuous family.

Proof. $|X_n^T(x_2) - X_n^T(x_1)|^2 = |\int_{x_1}^{x_2} \dot{X}_n^T(t)\, dt|^2$

$$\leq |x_2 - x_1|^2 \int_{x_1}^{x_2} |\dot{X}_n^T|^2\, dt$$

$$\leq 2|x_2-x_1|^2 \quad \int^{x_2} |\dot{X}_n^T - b(X_n^T)|^2\, ds + \int_0^T b(X_n^T)^2\, ds]$$

$$\leq 2|x_2 - x_1|^2\, [2M_T + T\, \|b\|_\infty^2].$$

Thus $\{X_n^T\}_n$ is an equicontinuous family. Since $\overline{G}$ is bounded, $\{X_n^T\}_n$ is uniformly bounded. By Arzela-Ascoli theorem and a "diagonal procedure" there exists a subsequence X_{n_k} and a continuous function to X uniformly on compact subsets of $[0,\infty)$. As $X_{n_k}(.) \in \overline{G}-D_1, X \in \overline{G}-D_1$

Let $m \geq n$. $I_0^{T_n}(X_m) \leq M$. $X_n \to X$ uniformly on $[0,T_n]$. By lower semi-continuity $I_0^T(X) \leq M$. Since this is true for every T we get on letting T tend to ∞, that

$$\frac{1}{2} \int_0^\infty |\dot{X} - b(X(s))|^2\, ds \leq M.$$

Thus we can find a sequence $a_1 < b_1 < a_2 < b_2 < \ldots$ such that

$$I_{a_n}^{b_n}(X(.)) = \frac{1}{2} \int_{a_n}^{b_n} |\dot{X}(t) - b(X(t))|^2\, dt$$

converges to zero with $b_n - a_n \to \infty$. Let $Y_n(t) = X(t + a_n)$. Then

$$I_0^{b_n - a_n}(Y_n) \to 0 \quad \text{with } b_n - a_n \to +\infty,\ Y_n \in \Omega_{\delta_1}.$$

Just as X was constructed from X_n, we can construct Y from Y_n such that $Y_n \to Y$ uniformly on compact subsets of $[0,\infty)$.

$$I_0^{b_n - a_n}(Y) \leq \inf_{m \geq n} I_0^{b_n - a_n}(Y_m) = 0$$

(by lower semi-confirmity of I_0^T). Thus $I_0^{b_n - a_n}(Y) = 0$, $\forall$ n, showing that

$$\int_0^{\infty} Y(t) - b(Y(t))|^2 \, dt = 0$$

Thus Y satisfies $Y(.) = x + \int_0^t b(Y(s))\, ds$ with $Y(t) \in \overline{G} - \partial D_1, \forall$ t.

<u>Case (i)</u>. $Y(t_0) \in G$ for some t_0. Let $Z(t) = Y(t + t_0)$ so that Z is an integral curve starting at a p oint of G and remaining away from D_1 contradicting the stability condition.

<u>Case (ii)</u>. $Y(t_0) \notin G$ for any t_0, i.e. $Y(t) \in \partial G$ for all t. Since $Y(t) = b(Y(t)) < Y(t), n(Y(t)) >$ is strictly positive. But $Y(t) \in \partial G$ and hence $< Y(t), n(Y(t)) > = 0$ which leads to a contradiction. Thus our assumption is incorrect and hence the lemma follows.

<u>LEMMA</u>. *Let* $x \in \partial D_2$ *and define*

E = {X(t) *exits from* G *before hitting* D_1 *and it exits from* N}

F = {X(t) *exits from* G *before hitting* D_1 *and it exits from* N^c}

Then

$$\frac{Q_x^{\varepsilon}(F)}{Q_x^{\varepsilon}(E)} \leq \exp\left(- \frac{3\eta}{8\varepsilon} + O\left(\frac{1}{\varepsilon}\right)\right) \to 0 \text{ uniformly in } x \ (x \in \partial D_2).$$

SIGNIFICANCE. $Q_x^\varepsilon(E)$ and $Q_x^\varepsilon(F)$ are both small because diffusion is small and the drift is large. The lemma says that $Q_x^\varepsilon(E)$ is relatively much larger than $Q_x^\varepsilon(F)$.

Proof. $Q_x^\varepsilon(E) \geq Q_x^\varepsilon$ $\{X(t)$ exits from G before hitting D_1, and exits in N before time $T\}$,

$= Q_x(B) \geq \exp[-\frac{1}{\varepsilon} \inf I_0^T(X(.))]$ where the infimum is taken over the interior of B,

$$\geq \exp[-\frac{1}{\varepsilon}(V + \frac{3\eta}{8}) + o(\frac{1}{\varepsilon})].$$

Similarly,

$$Q_x^\varepsilon(F) \leq \exp[-\frac{1}{\varepsilon}(V + \frac{3\eta}{4}) + o(\frac{1}{\varepsilon})].$$

Therefore

$$\frac{Q_x^\varepsilon(F)}{Q_x^\varepsilon(E)} \leq \exp[-\frac{3\eta}{8\varepsilon} + o(\frac{1}{\varepsilon})] \to 0 \quad \text{as} \quad \varepsilon \to 0.$$

We now proceed to prove the main theorem. Let

$$\tau_0 = 0,$$

$$\tau_1 = \text{first time } \partial D_1 \text{ is hit,}$$

$$\tau_2 = \text{next time } \partial D_2 \text{ is hit,}$$

$$\tau_3 = \text{next time } \partial D_1 \text{ is hit,}$$

....

and so on. Observe that the particle can get out of G only between the time intervals τ_{2n} and τ_{2n+1}. Let E_n = {between τ_{2n} and τ_{2n+1} the path exits from G for the first time and that it exits in $N\}$, $F_n = \{$between τ_{2n} and τ_{2n+1} the path exits from G for the first time and that it exits in $N^c\}$.

$$Q_x^\varepsilon(X(\tau) \in N) + Q_x(X(\tau) \in N^c) = 1.$$

Also

$$Q_x^\varepsilon(X(\tau) \in N^c) = \sum_{n=1}^{\infty} Q_x^\varepsilon(F_n),$$

$$Q_x^\varepsilon(X(\tau) \in N) = \sum_{n=1}^{\infty} Q_x^\varepsilon(E_n),$$

$$\sum_{n=1}^{\infty} Q_x^\varepsilon(F_n) = \sum_{n=1}^{\infty} E^{Q_x^\varepsilon}(Q_x^\varepsilon(F_n \mid \mathcal{F}_{\tau_{2n}}))$$

$$\leq \sum_{n=1} E^{Q_x^\varepsilon}[\chi_{(\tau > \tau_{2n})} \sup_{x \in \partial D_2} Q_x^\varepsilon(F)] \text{ (by the Strong Markov property)}$$

$$\leq O(\varepsilon) \sum_{n=1}^{\infty} E^{Q_x^\varepsilon}[\chi_{(\tau > \tau_{2n})} \inf_{x \in \partial D_2} Q_x^\varepsilon(E)] \quad \left(\text{as } \frac{Q_x^\varepsilon(F)}{Q_x^\varepsilon(E)} \to 0\right)$$

$$\leq O(\varepsilon) \sum_{n=1}^{\infty} Q_x^\varepsilon(E_n) = O(\varepsilon)\, Q_x(X(\tau) \in N^c).$$

Therefore

$$Q_x^\varepsilon(X(\tau) \in N) \to 1,\ Q_x(X(\tau) \in N^c) \to 0.$$

Exercise. Suppose $b(x) = \nabla u(x)$ for some $u \in C^1(G \cup \partial G, R)$. Assume that $u(x_0) = 0$ and $u(x) < 0$ for $x \neq x_0$. Show that

$$V(x_0, y) = -2u(y).$$

[Hint: For any trajectory X with $X(0) = x_0$,

$X(T) = y, I_0^T(X) = \frac{1}{2} \int_0^T |\dot{X} + \nabla u(X)|^2 dt - 2 \int_0^T \nabla u(X).\dot{X}(t)dt \geq -2u(y)$

so that $V(x_0,y) \geq -2u(y)$. For the other inequality, let u be a solution of $\dot{X}(t) + \nabla u(X(t)) = 0$ on $[0,\infty)$ with $X(0) = y$. Show that because $\frac{duX(s)}{ds}$ 0 for $X(s) \neq 0$ and x_0 is the only zero of u, $\lim_{t \to \infty} X(t) = x_0$. Now conclude that $V(x_0,y) \leq -u(y)$].

23. STOCHASTIC INTEGRAL FOR A WIDER CLASS OF FUNCTIONS

WE SHALL NOW define the stochastic integral for a wider class of functions.

Let $\theta: [0, \ ) X\Omega \to \mathbb{R}^d$ be any progressively measurable function such that for every t

$$\int_0^t |\theta(s,w)|^2 ds < \infty, \quad \text{a.e.}$$

Define, for every finite $L \geq 0$,

$$\theta_L(s,w) = \begin{cases} \theta(s,w), & \text{if } \int_0^s |\theta(t,w)|^2 dt < L < \infty, \\ 0 & , \text{if } \int_0^s |\theta(t,w)|^2 dt \geq L. \end{cases}$$

We can write $\theta_L(s,w) = \theta(s,w)\, \chi_{[0,L)}(\phi(s,w))$ where

$\phi(s,w) = \int_0^s |\theta(t,w)|^2 dt$ is progressively measurable. Hence

$\theta_L(s,w)$ is progressively measurable. It is clear that

$\int_0^T |\theta_L(s,w)|^2 ds \leq L$, a.e. $\forall$ T. Therefore $E(\int_0^T |\theta_L(s,w)|^2 ds) \leq L$.

Thus the stochastic integral $\xi_L(t,w) = \int_0^t < \theta_L(s,w) \ , dX(s,w) >$ is well defined.

The proofs of the next three lemmas follow closely the treatment of Stochastic integration given earlier.

LEMMA 1. *Let* τ *be a bounded, progressively measurable, continuous function. Let* τ *be any finite stopping time. If* $\theta(s,w) = 0, \forall (s,w)$

such that $0 \leq s \leq \tau(w)$ *then* $\int_0^t < \theta(s,w),\ dX(s,w) > = 0$ *for*

$0 \leq t \leq \tau(w)$.

Proof. Define $\theta_n(s,w) = \theta(\frac{[ns]}{n}, w)$. θ_n is progressively measurable and by definition of the stochastic integral of θ_n,

$$\int_0^t < \theta_n(s,w),\ dX(s,w) > = 0,\ \forall\ t,\ 0 \leq t \leq \tau(w).$$

Now

$$E(\int_0^t |\theta_n(s,w)-\theta(s,w)|^2 ds) = E(\int_0^t |\theta(\frac{[ns]}{n}\ w)-\theta(s,w)|^2 ds) \to 0 \text{ as } n \to \infty$$

and

$$\int_0^t < \theta_n(s,w),\ dX(s,w) > \to \int_0^t < \theta(s,w), dX(s,w) >$$

in probability. Therefore

$$\int_0^t < \theta,\ dX > = 0 \quad \text{if} \quad 0 \leq t \leq \tau(w).$$

LEMMA 2. *If* θ *is progressively measurable and bounded the assertion of Lemma* 1 *still holds.*

Proof. Let

$$\theta_n(t,w) = \frac{1}{n} \int_{(t - 1/n)V_0}^{t} \theta(s,w)\ ds.$$

Then

$$E(\int_0^T |\theta_n(t,w) - \theta(t,w)|^2 dt) \to 0 \quad \text{(Lebesgue's theorem).}$$

θ_n is continuous and bounded, $\theta_n(t,w) = 0$ for $0 \leq t \leq \tau(w)$. By lemma 1

$$\int_0^t < \theta_n(s,w), dX(s,w) > = 0$$

if $0 \leq t \leq \tau(w)$. This proves the result.

<u>LEMMA 3</u>. *Let* θ *be progressively measurable such that, for all* t,

$$E\left(\int_0^t |\theta(s,w)|^2 \, ds\right) < \infty.$$

If $\theta(s,w) = 0$ *for* $0 \leq s \leq \tau(w)$, *then*

$$\int_0^t < \theta(s,w), dX(s,w) > = 0 \quad \text{for} \quad 0 \leq t \leq \tau(w).$$

<u>Proof</u>. Define

$$\theta_n = \begin{cases} \theta, & \text{if } |\theta| \leq n, \\ 0, & \text{if } |\theta| > n. \end{cases}$$

Then

$$\int_0^t < \theta_n, dX > = 0, \text{ if } 0 \leq t \leq \tau(w), \text{ (Lemma 2) and}$$

$E\left(\int_0^t |\theta_n - \theta|^2 ds\right) \to 0$. The result follows.

<u>LEMMA 4</u>. *Let* θ *be progressively measurable such that* $\forall$ t,

$$\int_0^t |\theta(s,w)|^2 \, ds < \infty \quad \text{a.e.}$$

Then $\mathop{\mathrm{Lt}}_{L \to \infty} \xi_L(t,w)$ *exists a.e.*

Proof. Define

$$\tau_L(w) = \inf\left\{s: \int_0^s |\theta(\sigma,w)|^2\, d\sigma \geq L\right\};$$

clearly τ_L is a stopping time. If $L_1 \leq L_2$, $\tau_{L_1}(w) \leq \tau_{L_2}(w)$ and by assumptions of the lemma $\tau_L \uparrow \infty$ as $L \uparrow \infty$. If

$$L_1 \leq L_2,\ \theta_{L_1}(s,w) = \theta_{L_2}(s,w) \ \text{ for } \ 0 \leq s \leq \tau_{L_1}(w).$$

Therefore by Lemma 3,

$$\xi_{L_2}(t,w) = \xi_{L_1}(t,w)$$

if $0 \leq t \leq \tau_{L_1}(w)$. Therefore as soon as L is large enough such that $t \leq \tau_L(w)$, $\xi_L(t,w)$ remains constant (as a function of L). Therefore $\underset{L \to \infty}{\mathrm{Lt}}\ \xi_L(t,w)$ exists a.e.

DEFINITION. The stochastic integral of θ is defined by

$$\int_0^t \langle \theta(s,w),\ dX(s,w) \rangle = \underset{L \to \infty}{\mathrm{Lt}}\ \xi_L(t,w).$$

Exercise. Check that the definition of the stochastic integral given above coincides with the previous definition in case

$$E\left(\int_0^t |\theta(s,w)|^2 ds\right) < \infty,\ \forall\ t.$$

LEMMA. *Let* θ *be a progressively measurable function, such that*

$$\int_0^t |\theta(s,w)|^2\, ds < \infty,\ \forall\ t.$$

If $\xi(t,w)$ *denotes the stochastic integral of* θ, *then*

$$P\Big(\sup_{0 \le t \le T} |\xi\,(t,.)| \ge \varepsilon\Big) \le P\Big(\int_0^T |\theta|^2 ds \ge L\Big) + \frac{L^2}{\varepsilon^2} .$$

<u>Proof</u>. Let $\tau_L = \inf\{t: \int_0^t |\theta|^2 ds \ge L\}$. If $T < \tau_L(w)$, then $\theta_L(s,w) = \theta(s,w)$. Also

$$\xi_L(t,w) = \xi(t,w) \quad \text{for} \quad 0 \le t \le T.$$

<u>CLAIM</u>. $\{w: \sup_{0 \le t \le T} |\xi(t,w)| \ge \varepsilon\}$

$$\{w: \sup_{0 \le t \le T} |\xi_L(t,w)| \ge \varepsilon\} \cup \{w: \tau_L(w) \le T\}$$

For, if w is not contained in the right side, then

$$\sup_{0 \le t \le T} |\xi_L(t,w)|^2 < \varepsilon \text{ and } |\tau_L(w)| > T.$$

If $\tau_L > T$, $\xi_L(t,w) = \xi(t,w)\ \forall\ t \le T$. Therefore

$$\sup_{0 \le t \le T} |\xi_L(t,w)| = \sup_{0 \le t \le T} |\xi(t,w)|$$

Therefore $w \notin$ left side. Since

$$\{\ w: \tau_L(w) > T\} = \{w: \int_0^T |\theta|^2 ds \ge L\}$$

we get

$$P\Big(\sup_{0 \le t \le T} |\xi(t,.)| \ge \varepsilon\Big)$$

$$\le P\Big(\int_0^T |\theta|^2 ds \ge L\Big) + P\Big(\sup_{0 \le t \le T} |\xi_L(t,.)| \ge \varepsilon\Big)$$

$$\leq P\left(\int_0^T |\theta|^2 ds \geq L\right) + \frac{L^2}{\varepsilon^2},$$

by Kolmogorov's inequality. This proves the result.

COROLLARY. *Let* θ_n *and* θ *be progressively measurable functions such that*

(a) $$\int_0^t |\theta_n(s,w)|^2 \, ds < \infty, \quad \int_0^t |\theta(s,w)|^2 \, ds < \infty, \forall\, t;$$

(b) $$\operatorname*{Lt}_{n \to \infty} \int_0^t |\theta_n(s,w) - \theta(s,w)|^2 \, ds = 0 \quad \textit{in probability.}$$

If $\xi_n(t,w)$ *and* $\xi(t,w)$ *denote, respectively the stochastic integrals of* θ_n *and* θ*, then*

$$\sup_{0 \leq t \leq T} |\xi_n(t,w) - \xi(t,w)| \quad \textit{converges to zero in probability.}$$

Proof. Let $\tau_{n,L}(w) = \inf\left\{t : \int_0^t |\theta_n|^2 \, ds \geq L\right\}$; replacing θ by $\theta_n - \theta$ and ξ by $\xi_n - \xi$ in the previous lemma, we get

$$P\left(\sup_{0 \leq t \leq T} |\xi_n(t,.) - \xi(t,.)| \geq \varepsilon\right)$$

$$\leq \frac{L^2}{\varepsilon^2} + P\left(\int_0^T |\theta_n(s,.) - \theta(s,.)|^2 \, ds \geq L\right).$$

Letting $n \to \infty$, we get

$$\lim_{n \to \infty} P\left(\sup_{0 \leq t \leq T} |\xi_n(t,.) - \xi(t,.)| \geq \varepsilon\right) \quad \frac{L^2}{\varepsilon^2}.$$

As L is arbitrary we get the desired result.

PROPOSITION. *Let* θ *be progressively measurable such that*

$$\int_0^t |\theta(s,w)|^2 \, ds < \infty, \ \forall\, t \textit{ and } \forall\, w.$$

Then

$$(*) \qquad Z(t,.) = \exp\left[\int_0^t < \theta(s,.), dX(s,.) > - \frac{1}{2}\int_0^t |\theta(s,.)|^2 ds\right]$$

is a super martingale satisfying

(a) $E(Z(t,.)) \leq 1$;

(b) $\underset{t \to 0}{\text{Lt}}\ E(Z(t,.)) = 1.$

Proof. Let (θ_n) be a sequence of bounded progressively measurable functions such that

$$\int_0^t |\theta_n - \theta|^2 ds \to 0 \quad \forall\, t, \forall\, w.$$

(For example we may take $\theta_n = \theta$ if $|\theta| < n$, $= 0$ otherwise). Then (*) is a martingale when θ is replaced by θ_n. This martingale satisfies $E(Z_n(t,.)) = 1$, and $Z_n(t,.) \to Z(t,.)$ pointwise (a) now follows from Fatou's lemma:

$$\lim_{t \to 0} E(Z(t,.)) \geq E(\underline{\lim}_{t \to 0} Z(t,.))$$

$$\geq E(1) = 1.$$

Therefore $\underset{t \to 0}{\text{Lt}}\ E(Z(t,.)) = 1.$ This proves (b).

24. EXPLOSIONS

Exercise. Let $R > 0$ be given. Put $b_R = b\phi_R$ where $\phi_R = 1$ on $|x| \geq R$, $\phi_R = 0$ if $|x| \geq R + 1$; ϕ_R is C^∞. Show that $b_R = b$ on $|x| \leq R$, b_R is bounded on $\mathbb{R}^d$ and b_R is globally Lipschitz.

Let $\Omega_T = \{w \in \Omega: B(w) > T\}$. Let $S^T = \Omega_T \to C[0,T]$ be the map $S^T w = y(.)$ where $y(t) = w(t) + \int_0^t b(y(s))ds$ on $[0,T]$. Unless otherwise specified $b: \mathbb{R}^d \to \mathbb{R}^d$ is assumed to be locally Lipschitz. Define the measure Q_x^T on $(\Omega, {}_T)$ by

$$Q_x^T(A) = P_x\{w: S^T w \in A, B(w) > T\},$$

where P_x is the probability measure corresponding to Brownian motion.

THEOREM. $$Q_x^T(A) = \int_A Z(T,.)dP_x, \ \forall A \in \mathcal{F}_T,$$

where

$$Z(T,.) = \exp[\int_0^T \langle b, dX \rangle - \frac{1}{2}\int_0^T |b(X(s,.))|^2 ds].$$

REMARK. If b is bounded or if b satisfies a global Lipschitz condition then $B(w) = \infty$, so that $\Omega_T = \Omega$ and Q_x^T are probability measures.

Proof. Let $0 \leq R < \infty$. For any w in Ω, let y be given by

$$y(t) = w(t) + \int_0^t b(y(\sigma))d\sigma.$$

Define $\sigma_R(w) = \inf\{t: |y(t)| \geq R$ and let b_R be as in the Exercise. Then the equation

$$y_R(t) = w(t) + \int_0^t b_R(y_R(\sigma))d\sigma$$

has a global solution. Denote by $S_R: \Omega \to \Omega$ the map $w \to y_R$. If $Q_{R,x}$ is the measure induced by S_R, then

$$\frac{dQ_{R,x}}{dP_x}\bigg|_{\mathcal{F}_t} = Z_R(t) = \exp\left(\int_0^t \langle b_R, dX\rangle - \frac{1}{2}\int_0^t |b_R|^2 ds\right).$$

Let $\tau_R(w) = \inf\{t: |w(t)| > R\}$. τ_R is a stopping time satisfying $\tau_R S_R = \sigma_R$. By the optional stopping theorem,

(1) $$\frac{dQ_{R,x}}{dP_x}\bigg|_{\mathcal{F}_{\tau_R \wedge T}} = Z_R(\tau_R \wedge T) = Z(\tau_R \wedge T).$$

<u>CLAIM</u>. $Q_{R,x}((\tau_R > T) \cap A) = Q_x^T((\tau_R > T) \cap A)$, $\forall$ A in $\mathcal{F}_T$.

<u>Proof</u>. Right side $= P_x\{w: B(w) > T, S^T(w) \in A \cap (\tau_R > T)\}$

$= P_x\{w: B(w) > T, y \in A, \sup_{0 \le t \le T} |y(t)| < R\}$

$= P_x\{w: y$ is defined at least upto time T,

$y \in A, \sup_{0 \le t \le T} |y(t)| > R\}$

$= P_x\{w: y_R \in A, \sup_{0 \le t \le T} |y_R(t)| < R\}$

$= P_x\{w: S_R(w) \in A, \tau_R S_R(w) > T\}$

$= Q_{R,x}\{(\tau_R > T) \cap A\}$

(by definition). As Ω is an increasing union of $\{\tau_R > T\}$ for R increasing,

$$Q_x^T(A) = \lim_{R \to +\infty} Q_x^T((\tau_R > T) \cap A), \quad \forall A \text{ in } \mathcal{F}_T,$$

$$= \lim_{R \to \infty} Q_{R,x}((\tau_R > T) \cap A) \quad \text{(by claim)}$$

$$= \lim_{R \to \infty} \int_{(\tau_R \wedge T) \cap A} \exp\left(\int_0^{\tau_R \wedge T} \langle b, dX \rangle - \frac{1}{2}\int_0^{\tau_R \wedge T} |b|^2 ds\right) dP_x \quad \text{(by (1))}$$

$$= \int_A \exp\left(\int_0^T \langle b, dX \rangle - \frac{1}{2}\int_0^T |b|^2 ds\right) dP_x$$

$$= \int_A Z(T)\, dP_x.$$

THEOREM. *Suppose* $b: \mathbb{R}^d \to \mathbb{R}^d$ *is locally Lipschitz; let* $L = \frac{\Delta}{2} + b.\nabla$.

(i) *If there exists a smooth function* $u: \mathbb{R}^d \to (0,\infty)$ *such that* $u(x) \to \infty$ *as* $|x| \to \infty$ *and* $Lu \leq cu$ *for some* $c > 0$ *then* $P_x\{w: B(w) < \infty\} = 0$, *i.e. for almost all* w *there is no explosion.*

(ii) *If there exists a smooth bounded function* $u: \mathbb{R}^d \to (0,\infty)$ *such that* $Lu \geq cu$ *for some* $c > 0$, *then* $P_x\{w: B(w) < \infty\} > 0$, *i.e. there is explosion.*

CORROLLARY. *Suppose, in particular,* b *satisfies* $|\langle b(x),x \rangle| \leq A+B|x|^2$ *for some constants* A *and* B; *then* $P_x(w: B(w) < \infty) = 0$.

Proof. Take $u(x) = 1 + |x|^2$ and use part (1) of the theorem..

Proof of theorem. Let b_R be as in the Exercise and let $L_R = \frac{\Delta}{2} + b_R.\nabla$; then $L_R u(x) \leq cu(x)$ if $|x| \leq R$.

CLAIM. $u(X(t))\, e^{-ct}$ is a supermartingale upto time τ_R relative to Q_x^R,

$$d(u(X(t))e^{-ct}\exp(\int_0^t <b_R,dX> - \frac{1}{2}\int_0^t |b_R|^2\,ds))$$

$$= e^{-ct}\, e^{\int_0^t <b_R,dX> - \frac{1}{2}\int_0^t |b_R|^2 ds} \times$$

$$\times \{-cudt + <\nabla u,dX> + u(x)[<b_R,dX> - \frac{|b_R|^2}{2}dt] + b_R u\,dt + \frac{1}{2}|b_R|^2 u\,dt\}$$

$$= \exp(-ct + \int_0^t <b_R,dX> - \frac{1}{2}\int_0^t |b_R|^2 ds).[L_R - c)u + <\nabla u,dX> + u<b_R,dX>].$$

Therefore

$$u(X(t))e^{-ct}\, e^{\int_0^t <b_R,dX> - \frac{1}{2}\int_0^t |b_R|^2 ds}$$

$$-\int_0^t \exp(-cs + \int_0^s <b_R,dX> - \int_0^s |b_R|^2 ds).(L_R - c)u(X(s))ds$$

is a Brownian stochastic integral. Therefore

$$u(X(\tau_R \wedge t))\exp(-c(\tau_R \wedge t) + \int_0^{\tau_R \wedge t} <b_R,dX> - \frac{1}{2}\int_0^{\tau_R \wedge t} |b_R|^2 ds) -$$

$$-\int_0^{\tau_R \wedge t} \exp(-cs + \int_0^s <b_R,dX> - \frac{1}{2}\int_0^s |b_R|^2 ds)\,(L_R - c)u(X(s))ds$$

is a martingale relative to P_x, $\mathcal{F}_{\tau_R \wedge t}$. But $b_R(x) = b(x)$ if $|x| \le R$. Therefore

$$u(X(\tau_R \wedge t))\exp(-c(\tau_R \wedge t) + \int_0^{\tau_R \wedge t} <b,dX> - \frac{1}{2}\int_0^{\tau_R \wedge t} |b|^2 ds) -$$

$$-\int_0^{\tau_R \wedge t} \exp(-cs + \int^s <b,dX> - \int_0^s |b|^2 ds)(L_R - c)u(X(s))ds$$

is a martingale relative to $\mathcal{F}_{\tau_R \wedge t}$. But $(L_R - c)u \le 0$ in $[0,\tau_R]$.

Therefore

$$u(X(\tau_R \wedge t))\exp\left(-c(\tau_R \wedge t)+\int_0^{\tau_R \wedge t} \langle b,dX \rangle - \frac{1}{2}\int_0^{\tau_R \wedge t} |b|^2 ds\right)$$

is a supermartingale relative to $\mathcal{F}_{\tau_R \wedge t}, P_x$. Therefore

$u(X(\tau_R \wedge t)\, e^{-c(\tau_R \wedge t)}$ is a supermartingale relative to Q_x^R

(optional sampling theorem). Therefore

$$E^{Q_x^R}(u(X(t_R \wedge t))e^{-c(\tau_R \wedge t)} \leq u(x);$$

letting $t \to \infty$, we get, using Fatou's lemma,

$$E^{Q_x^R}(u(X(\tau_R)\, e^{-c\tau_R}) \leq u(x).$$

Therefore

$$E^{Q_x^R}(e^{-c\tau_R}) \leq \frac{u(x)}{\inf\limits_{|y| = R} |u(y)|} .$$

Thus

$$E^{P_x}(e^{-c\sigma_R}) \leq \frac{u(x)}{\inf\limits_{|y| = R} |u(y)|}$$

(by change of variable). Let $R \to \infty$ to get $\underset{R \to \infty}{\text{Lt}} \int e^{-c\sigma_R}\, dP_x = 0$,

i.e. $P_x\{w: B(w) < \infty\} = 0$.

Sketch of proof for Part (ii).

By using the same technique as for Part (i), show that $u(X(t))e^{-ct}$ is a submartingale upto time τ_R relative to Q_x^R, so that

$$E^{P_x}\left(e^{-c\sigma_R}\right) \geq \frac{u(x)}{\sup\limits_{|y| = R} |u(y)|} \geq \frac{u(x)}{\|u\|_\infty} > 0 ;$$

let $R \to \infty$ to get the result.

<u>Exercise</u>. Show that if $L = \frac{1}{2}\frac{\partial^2}{\partial x} + x^3 \frac{\partial}{\partial x}$, there is explosion.

(Hint: take $u = e^{\tan^{-1}(x^2)}$ and show that $Lu \geq u$).

25. CONSTRUCTION OF A DIFFUSION PROCESS

Problem. Given $a: [0,\infty) \times \mathbb{R}^d \to S_d^+$, bounded measurable and $b: [0,\infty) \times \mathbb{R}^d \to \mathbb{R}^d$ bounded measurable, to find $(\Omega, \mathcal{F}_t, P, X)$ where Ω is a space, $\{\mathcal{F}_t\}_{t \geq 0}$ an increasing family of σ-algebras on Ω, P a probability measure on the smallest σ-algeb ra containing all the $\mathcal{F}_t$'s. and $X: [0,t) \times \Omega \to \mathbb{R}^d$, a progressively measurable function such that $X(t,w) \in I[b(t,X_t), a(t,X_t)]$.

Let $\Omega = C[0,\infty); \mathbb{R}^n)$, $\beta(t,.)$ = n-dimensional Brownian motion, $\mathcal{F}_t = \sigma\{\beta(s): 0 \leq s \leq t\}$, P the Brownian measure on Ω and a and b as given in the problem. We shall show that thee problem has a solution, under some special conditions on a and b.

THEOREM. *Assume that there exists* $\sigma: [0,\infty) \times \mathbb{R}^d \to M_{d \times n}$ ($M_{d \times n}$ = *set of all* $d \times n$ *matrices over the reals) such that* $\sigma\sigma^* = a$. *Further let*

$$\sum_{i,j} |\sigma_{ij}(t,x)| \leq C, \quad \sum_j |b_j(t,x)| \leq C,$$

$$\sum_{i,j} |\sigma_{ij}(t,x_1) - \sigma_{ij}(t,x_2)| \leq A|x_1 - x_2|,$$

$$\sum_j |b_j(t,x_1) - b_j(t,x_2)| \leq A|x_1 - x_2|.$$

Then the equation

$$(1) \quad \xi(t,.) = x + \int_0^t <\sigma(s,\xi(s,.)), d\beta(s,.)> + \int_0^t b(s,\xi(s,.))\, ds$$

has a solution. The solution $\xi(t,w): [0,\infty) \times \Omega \to \mathbb{R}^d$ *can be taken to be such that* $\xi(t,.)$ *is progressively measurable and such that* $\xi(t,.)$ *is continuous for* a, *a.e. If* ξ, η *are progressively measurable, continuous (for* a.a.e) *solutions of equation (1), then* $\xi = \eta$ *for* a.a.w.

<u>Proof</u>. The proof proceeds in several steps.

<u>LEMMA 1</u>. *Let* Ω *be any space with* $(\mathcal{F}_t)_{t \geq 0}$ *an increasing family of* σ*-algebras. If* $0 \leq T \leq \infty$ *then there exists a* σ*-algebra* $\mathcal{A}_0 \subset \mathcal{A} = \mathcal{B}[0,T] \times \mathcal{F}_T$ *such that a function* $f: [0,T] \times \Omega \to \mathbb{R}$ *is progressively measurable if and only if* f *is measurable with respect to* $\mathcal{A}_0$.

<u>Proof</u>. Let $\mathcal{A}_0 = \{A \in \mathcal{A} : \chi_A$ is progressively measurable$\}$. Clearly $[0,T] \times \Omega \in \mathcal{A}_0$, and if $A \in \mathcal{A}_0$, $A^c \in \mathcal{A}_0$. Thus $\mathcal{A}_0$ is an algebra. As increasing limits (decreasing limits) of progressively measurable functions are progressively measurable, $\mathcal{A}_0$ is a monotone class and hence a σ-algebra.

Let $f: [0,T] \times \Omega \to \mathbb{R}$ be progressively measurable; in fact, $f^+ = \frac{f + 1 = f|}{2}$, $f^- = \frac{f - |f|}{2}$. Let $g = f^+$. Then

$$g_n = \sum_{i=1}^{n2^n} \frac{i-1}{2^n} \chi_{g^{-1}[\frac{i-1}{2^n}, \frac{i}{2^n})} + n \chi_{g^{-1}[n,)}$$

is progressively measurable. Hence $n \vee g_n$ is progressively measurable, i.e. $n\chi_{g^{-1}[n,\infty)}$ is progressively measurable. Similarly

$\chi_{g^{-1}[\frac{i-1}{2^n}, \frac{i}{2^n})}$ is progressively measurable, etc. Therefore, by definition, g_n is measurable with respect to $\mathcal{A}_0$. As $g = f^+$ is the pointwise limit of g_n, f^+ is meas urable with respect to $\mathcal{A}_0$. Similarly f^- is $\mathcal{A}_0$-measurable. Thus f is $\mathcal{A}_0$-measurable.

Let $f: [0,T] \times \Omega \to \mathbb{R}$ be measurable with respect to $\mathcal{A}_0$. Again, if $g = f^+$

$$g_n = \sum_{i=1}^{n2^n} \frac{i-1}{2^n} \chi_{g^{-1}[\frac{i-1}{2^n}, \frac{i}{2^n})} + n\chi_{g^{-1}[n,\infty)}$$

is $\mathcal{A}_0$-measurable. Since $g^{-1}[n,\infty), \ldots g^{-1}[\frac{i-1}{2^n}, \frac{i}{2^n}) \in \mathcal{A}_0$. So g_n is progressively measurable. Therefore g is progressively measurable. Hence f is progressively measurable. This completes the proof of the Lemma.

To solve (1) we use the standard iteration technique.

<u>Step 1</u>. Let $\xi_0(t,w) = x$,

$$\xi_n(t,w) = x + \int_0^t < \sigma(s,\xi_{n-1}(s,w)), d\beta(s,w) > + \int_0^t b(s,\xi_{n-1}(s,w))ds.$$

By induction, it follows that $\xi_n(t,w)$ is progressively measurable.

<u>Step 2</u>. Let $\Delta_n(t) = E(|\xi_{n+1}(t) - \xi_n(t)|^2)$. If $0 \leq t \leq T, \Delta_n(t) \leq C^* \int_0^t \Delta_{n-1}(s)\, ds$ and $\Delta_0(t) \leq C^*t$, where C^* is a cons tant depending only on T.

Proof. $\Delta_0(t) = E(|\xi(t) - x|^2)$

$$= E(|\int_0^t < \sigma(s,x), d\beta(s,x) > + \int_0^t b(s,x)\, ds|^2)$$

$$\leq 2E(|\int_0^t < \sigma(s,x), d\beta(s,x) > |^2) +$$

$$+ 2E(|\int_0^t b(s,x) ds|^2) \text{ (use the fact that } |x+y|^2$$

$$\leq 2(|x|^2 + |y|^2) \ \forall\ x,y \in \mathbb{R}^d)$$

$$= 2E(\int_0^t \operatorname{Tr} \sigma\sigma^* ds) + 2E(|\int_0^t b(s,x) ds|^2).$$

or

$$\Delta_0(t) \leq 2E(\int_0^t \operatorname{tr} \sigma\sigma^* ds) + 2E(t \int_0^t |b(s,x)|^2 ds)$$

(Cauchy-Schwarz inequality)

$$\leq 2nd\, C^2(1 + T)t.$$

$$\Delta_n(t) = E(|\xi_{n+1}(t) - \xi_n(t)|^2)$$

$$= E(|\int_0^t < \sigma(s,\xi_n(s,w)) - \sigma(s,\xi_{n-1}(s,w)) d\beta > +$$

$$+ \int_0^t b(s,\xi_n(s,w)) - b(s,\xi_{n-1}(s,w)) ds|^2)$$

$$\leq 2E(|\int_0^t < \sigma(s,\xi_n(s,w)) - (s,\xi_{n-1}(s,w)), d\beta(s,w) > |^2) +$$

$$+ 2E(|\int_0^t (b(s,\xi_n(s,w)) - b(s,\xi_{n-1}(s,w)) ds|^2)$$

$$\leq 2E(\int_0^t \operatorname{tr} [(\sigma(s,\xi_n(s,w)) - \sigma(s,\xi_{n-1}(s,w))] \times$$

$$\times [\sigma^*(s,\xi_n(s,w)) - \sigma^*(s,\xi_{n-1}(s,w))]\, ds] +$$

$$+ 2E\left(t \int_0^t |b(s,\xi_n(s,w)) - b(s,\xi_{n-1}(s,w))|^2 ds\right)$$

$$\leq 2\, dn\, A^2 \int_0^t \Delta_{n-1}(s)\, ds + 2\, t\, A^2 n \int_0^t \Delta_{n-1}(s)\, ds$$

$$\leq 2dn\, A^2(1 + T) \int_0^t \Delta_{n-1}(s)\, ds.$$

This proves the result.

<u>Step 3</u>. $\Delta_n(t) \leq \frac{(C^*t)^{n+1}}{(n+1)!}$ $\forall\, n$ in $0 \leq t \leq T$, where

$$C^* = \max\left\{2nd\, C^2(1 + T),\ wnd\, A^2(1 + T)\right\}.$$

Proof follows by induction on n.

<u>Step 4</u>. $\xi_n|_{[0,T] \times \Omega}$ is Cauchy in $L^2([0,T] \times \Omega, B([0,T]\times \Omega),$ $\mu \times P)$, where μ is the Lebesgue measure on $[0,T]$.

<u>Proof</u>. $\Delta_n(t) \leq \frac{(C^*t)^{n+1}}{(n+1)!}$ implies that

$$\|\xi_{n+1} - \xi_n\|_2^2 \leq \frac{(C^*T)^{n+2}}{(n+2)!}.$$

Here $\|\cdot\|_2$ is the norm in $L^2([0,T] \times \Omega)$. Thus

$$\sum_{n=1}^{\infty} \|\xi_{n+1} - \xi_n\|_2 < \infty, \text{ proving Step (4).}$$

<u>Step 5</u>. (4) implies that $\xi_n|_{[0,T] \times \Omega}$ is Cauchy in $L^2([0,T] \times \Omega, \mathcal{A}_0, \mu \times P)$ where $\mathcal{A}_0$ is as in Lemma 1. Thus $\xi_n|_{[0,T] \times \Omega}$ converges to $\overline{\xi}_T$ in $L^2([0,T] \times \Omega)$ where $\overline{\xi}_T$ is progressively measurable.

Step 6. If $\xi_n|_{[0,T_2]\times\Omega} \to \overline{\xi}_{T_2}$ in $L^2([0,T_2]\times\Omega)$ and

$$\xi_n|_{[0,T_1]\times\Omega} \to \overline{\xi}_{T_1} \text{ in } L^2([0,T_1]\times\Omega),$$

then $\overline{\xi}_{T_2}|_{[0,T_1]\times\Omega} = \overline{\xi}_{T_1}$ a.e. on $[0,T_1]\times\Omega$, $T_1 < T_2$.

This follows from the fact that if $\xi_n \to \xi$ in L^2, a subsequence of (ξ_n) converges pointwise a.e. to ξ.

Step 7. Let $\overline{\xi}$ be defined on $[0,\infty)\times\Omega$ by $\overline{\xi}|_{[0,T]\times\Omega} = \overline{\xi}_T$. We now show that

$$\overline{\xi}(t,w) = x + \int_0^t \langle \sigma(s,\overline{\xi}(s,.)), d\beta(s,.)\rangle + \int_0^t b(s,\overline{\xi}(s,.))ds.$$

Proof. Let $0 \le t \le T$. By definition,

$$\xi_n(t,w) = x + \int_0^t \langle \sigma(s,\xi_{n-1}(s,.)), d\beta(s,.)\rangle + \int_0^t b(s,\xi_{n-1}(s,.))ds.$$

$$E[(\int_0^t \langle (\sigma(s,\xi_n(s,.)) - \sigma(s,\overline{\xi}(s,w))), d\beta(s,w)\rangle)^2]$$

$$= E(\int_0^t tr[(\sigma(s,\xi_n(s,w)) - \sigma(s,\overline{\xi}(s,w)))(\sigma(s,\xi_n(s,w)) - \sigma(s,\overline{\xi}(s,w)))^* ds$$

$$\le dn\, A \int_0^T \int_\Omega |\xi_n(s,w) - \overline{\xi}(s,w)|^2 ds \to 0 \quad \text{as} \quad n\to\infty$$

(by Lipschitz condition on σ).

Therefore

$$\int_0^t \langle \sigma(s,\xi_{n-1}(s,w), d\beta(s,w)\rangle \to \int_0^t \langle \sigma(s,\overline{\xi}(s,w)), d\beta(s,w)\rangle$$

in $L^2(\Omega, \ ,P)$. Similarly,

$$\int_0^t b(s,\xi_n(s,w))ds \to \int_0^t b(s,\overline{\xi}(s,w))ds, \quad \text{in } L^2.$$

Thus we get

$$(*) \qquad \overline{\xi}(t,w) = x + \int_0^t < \sigma(s,\overline{\xi}(s,w)),d\beta(s,w) > +$$

$$+ \int_0^t b(s,\overline{\xi}(s,w))ds \quad \text{a.e. in } t, w.$$

Step 8. Let $\xi(t,w) \equiv$ the right hand side of (*) above. Then $\xi(t,w)$ is almost surely continuous because the stochastic integral of a bounded progressively measurable function is almost surely continuous. The result follows by noting that $[0,\infty) = \bigcup_{n=1}^{\infty} [0,n]$ and a function on $[0,\infty)$ is continuous iff it is continuous on $[0,n]$, $\forall$ n.

Step 9. Replacing $\overline{\xi}$ by ξ in the right side of (*) we get a solution

$$\xi(t,w) = x + \int_0^t < \sigma(s,\xi),d\beta > + \int_0^t b(s,\xi(s \ ,w))ds$$

that is a.s. continuous $\forall$ t and a.e.

Uniqueness. Let ξ and η be two progressively measurable a.s. continuous functions satisfying (1). As in Step 3,

$$E(|\xi(t,w)-x|^2) \leq 2(E(\int_0^t tr\ \sigma\sigma^* ds) + 2E(t \int_0^t b|^2 ds)$$

$$\leq 2E\left(\int_0^T \operatorname{tr} \sigma\sigma^* \, ds\right) + 2E\left(T\int_0^T |b|^2 ds\right), \text{ if } 0 \leq t \leq T$$

$$< \infty .$$

Thus $E(|\xi(t,w)|^2)$ is bounded in $0 \leq t \leq T$. Therefore

$$\phi(t) = E(|\xi(t,w) - \eta(t,w)|^2)$$

$$\leq 2E(|\xi(t,w)|^2) + 2E(|\eta(t,w)|^2)$$

and so $\phi(t)$ is bounded in $0 \leq t \leq T$. But

$$\phi(t) \leq 2dn A^2(1 + T) \int_0^t \phi(s) \, ds$$

as in Step 2; using boundedness of $\phi(t)$ in $0 \leq t \leq T$ we can find a constant C such that

$$\phi(t) \leq Ct \text{ and } \phi(t) \leq C \int_0^t \phi(s)ds, \; 0 \leq t \leq T.$$

By iteration $\phi(t) \leq \frac{(Ct)^n}{n!} \leq \frac{(CT)^n}{n!}$. Therefore

$$\phi = 0 \text{ on } [0,T],$$

i.e. $\xi(t,w) = \eta(t,w)$ a.e. in $[0,T]$. But rationals being dense in $\mathbb{R}$ we have

$$\xi = \eta \text{ a.e. and } \forall t.$$

It is now clear that $\xi \in I[b,a]$.

<u>REMARK</u>. The above theorem is valid for the equation

$$\xi(t,w) = x_0 + \int_{t_0}^t \langle \sigma(s, \;), d\beta \rangle + \int_{t_0}^t b(s,\xi)ds, \; \forall t \geq t_0.$$

This solution will be denoted by ξ_{t_0,x_0}.

PROPOSITION. *Let* $\phi: C([0,\infty); \mathbb{R}^n) \to C([t_0,\infty); \mathbb{R}^d)$ *be the map sending* w *to* ξ_{t_0,x_0}, P *the Brownian measure on* $C([0,\infty); \mathbb{R}^n)$. *Let* $P_{t_0,x_0} = P\phi^{-1}$ *be the measure induced on* $C([t_0,\infty); \mathbb{R}^d)$. *Define* $X: [t_0,\infty) \times C)[t_0,\infty); \mathbb{R}^d)$ *by* $X(t,w) = w(t)$. *Then* X *is an Itö process relative to* $(C([t_0,\infty); \mathbb{R}^d), \quad _{t_0}, P_{t_0,x_0})$ *with parameters*

$$[b(t,X_t), a(t,X_t)].$$

The proof of the proposition follows from

Exercise. Let $(\overline{\Omega}, \overline{\mathcal{F}_t}, \overline{P})$, $(\Omega, \mathcal{F}_t, P)$ be any two measure spaces with X,Y progressively measurable on Ω, $\overline{\Omega}$ respectively. Suppose $\lambda: \overline{\Omega} \to \Omega$ is such that λ is $(\overline{\mathcal{F}_t}, \mathcal{F}_t)$-measurable for all t, and $\overline{P}\lambda^{-1} = P$. Let $X(t,\overline{w}) = Y(t,\lambda w)$, $\forall\, \overline{w} \in \overline{\Omega}$. Show that

(a) If X is a martingale, so is Y.

(b) If $X \in I[b(t,X_t), a(t,X_t)]$ then

$Y \in I[b(t,Y_t), a(t,Y_t)]$.

LEMMA. *Let* $f: \mathbb{R}^2 \to \mathbb{R}$ *be* $(\Omega, \ , P)$ *-measurable,* $\sum$ *a sub-σ-algebra of* $\mathcal{F}$. *Let* $X: \Omega \to \mathbb{R}$ *and* $Y: \Omega \to \mathbb{R}$ *be such that* X *is* $\sum$*-measurable and* Y *is* $\sum$*-independent. If* $g(w) = f(X(w), Y(w))$ *with* $E(g(w)) < \infty$, *then*

$$E(g|\textstyle\sum)(w) = E(f(x,Y)|_{x\,=\,X(w)},$$

i.e.

$$E(f(X,Y)|_{\Sigma})(w) = \int_{\Omega} f(X(w), Y(w'))\, dP(w').$$

<u>Proof</u>. Let A and B be measurable subsets in $\mathbb{R}$. The result is trivially verified if $f = \chi_{A \times B}$. The set

$$\mathcal{A} = \{F \in \mathbb{R} : \text{the result is true for } \chi_F\}$$

is a monotone class containing all measurable rectangles. Thus the Lemma is true for all characteristic functions. The general result follows by limiting procedures.

26. UNIQUENESS OF DIFFUSION PROCESS

IN THE LAST section we proved that

$$\xi(t,w) = x_0 + \int_{t_0}^{t} < \sigma(s,\xi(s,w)), d\beta(s\ ,w) \ + \int_{t_0}^{t} b(s,\ (s,w)ds$$

has a solution under certain conditions on b and σ where $\sigma\sigma^* = a$. The measure $P_{t_0,x_0} = P\xi^{-1}_{t_0,x_0}$ was constructed on $(C([t_0,\infty);\ \mathbb{R}^d),\ \mathcal{F}_{t_0})$ so that the map $X(t,w) = w(\ t)$ is an Itô process with parameters b and a. We now settle the uniqueness question, about the diffusion process.

THEOREM. *Let*

(i) $a: [0,\infty) \times \mathbb{R}^d \to S_d^+$ *and* $b: [0,\infty) \times \mathbb{R}^d \to \mathbb{R}^d$ *be bounded measurable functions;*

(ii) $\Omega = C([0,\ );\ \mathbb{R}^d)$;

(iii) $X: [0,\infty) \times \Omega \to \mathbb{R}^d$ *be defined by* $X(t,w) = w(t)$;

(iv) $_t = \sigma\{X(s): 0 \leq s \leq t\}$;

(v) P *be any probability measure on*

$$= \sigma(\bigcup_{t \geq 0}\ _t)$$

such that $P\{X(0) = x_0\} = 1$ *and* X *is an Itô process relative to* $(\Omega,\ _t, P)$ *with parameters* $b(t,X_t)$ *and* $a(t,X_t)$;

(vi) $\sigma: [0,\infty) \times \mathbb{R}^d \to M_{dxn}$ *be a bounded measurable map into the set of all real dxn matrices such that* $\sigma\sigma^* = a$ *on* $[0,\infty) \times \mathbb{R}^d$.

Then there exists a generalised n *-dimensional Brownian motion* β *on* $(\bar{\Omega},\Sigma_t,Q)$ *and a progressively measurable a.s. continuous map* $\xi\colon [0,\infty)\times\bar{\Omega}\to\mathbb{R}^d$ *satisfying the equation*

$$(1)\qquad \xi(t,\bar{w}) = x_0 + \int_0^t <\sigma(s,\xi(s,\bar{w})),d\beta(s,\bar{w})> + \int_0^t b(s,\xi(s,\bar{w}))\,ds$$

with $Q\xi^{-1} = P$, *where* $\xi\colon\bar{\Omega}\to\Omega$ *is given by* $(\xi(\bar{w}))(t) = \xi(t,\bar{w})$.

Roughly speaking, any Itô process can be realised by means of a diffusion process governed by equation (1) with $\sigma\sigma^* = a$.

Proof., Case (i). Assume that there exist constants $m,M > 0$ such that $mI \le a(t,x) \le MI$ and σ is a $d\times d$ matrix satisfying $\sigma\sigma^* = a$. In this case we can identify $(\bar{\Omega},\Sigma_t,Q)$ with $(\Omega,\mathcal{F}_t,P)$. Since $D(t,.)$ is an Itô process,

$$\exp <\theta,X(t)> - \int_0^t <\theta,b(s,X(s,.))> ds - \frac{1}{2}\int_0^t <\theta,a(s,X(s,.))\theta> ds$$

is a $(\Omega,\mathcal{F}_t,P)$ -martingale. Put

$$Y(t,w) = X(t,w) - \int_0^t b(s,X(s,w))ds - x_0.$$

Clearly $Y(t,w)$ is an Itô process corresponding to the parameters $[0,a(s,X_s)]$, so that

$$\exp <\theta,Y(t,w)> - \frac{1}{2}\int_0^t <\theta,a(s,X(s,.))\theta> ds$$

is a $(\Omega,\mathcal{F}_t,P)$ -martingale. The conditions $m \le a \le M$ imply that σ^{-1} exists and is bounded. Let

$$\eta(t) = \int_0^t \sigma^{-1}\, dY = \int_0^t \sigma^{-1}(s,X(s,.))dY(s,.),$$

so that (by definition of a stochastic integral) η is a $(\Omega, \mathcal{F}_t, P)$ - Itô process with parameters zero and σ^{-1} a $(\sigma^{-1})^* = 1$. Thus η is a Brownian motion relative to $(\Omega, \mathcal{F}_t, P)$. Now by change of variable formula for stochastic integrals,

$$\int_0^t \sigma d\eta = \int_0^t \sigma\sigma^{-1}\, dY$$

$$= Y(t) - Y(0) = Y((t),$$

since $Y(0) = 0$. Thus

$$X(t) = x_0 + \int_0^t \sigma(s,X(s,.))d + \int_0^t b(s,X(s,.))ds.$$

Taking $Q = P$ we get the result.

Case (ii). $a = 0$, $b = 0$, $x_0 = 0$, $\sigma = 0$ where $\sigma \in M_{dxn}$. Let $(\Omega^*, \mathcal{F}_t^*, P^*)$ be an n -dimensional Brownian motion. Define

$$(\bar{\Omega}, \Sigma_t, Q) = (\Omega \times \Omega^*, \mathcal{F}_t \times \mathcal{F}_t^*, P \times P^*) .$$

If β is the n -dimensional Brownian motion on $(\Omega^*, \mathcal{F}_t^*, P^*)$, we define $\bar{\beta}$ on $\bar{\Omega}$ by $\bar{\beta}(t,w,w^*) = \beta(t,w^*)$. It is easy to verify that $\bar{\beta}$ is an n -dimensional Brownian motion on $(\bar{\Omega}, \Sigma_t, Q)$. Taking $\xi(t,w,w^*) = x_0$ we get the result.

Before we take up the general case we prove a few Lemmas.

LEMMA 1. *Let* $\sigma: \mathbb{R}^n \to \mathbb{R}^d$ *be linear* $\sigma\sigma^* = a: \mathbb{R}^d \to \mathbb{R}^d$; *then there exists a linear map which we denote by* $\sigma^{-1}: \mathbb{R}^d \to \mathbb{R}^n$ *ssuch that* $\sigma^{-1} a \sigma^{-1*} = \pi_{N_\sigma^\perp}$, *where* π *denotes the projection and* N_σ *null space of* σ.

Proof. Let R_σ = range of σ. Clearly $\sigma: N_\sigma^\perp \to R$ is an isomorphism. Let $\tau: R_\sigma \to N_\sigma^\perp$ be the inverse. We put

$$\sigma^{-1} = \tau \oplus 0: R_\sigma \oplus R_\sigma^\perp \to N_\sigma^\perp \oplus N_\sigma.$$

LEMMA 2. *Let* X, Y *be martingales relative to* $(\Omega, \mathcal{F}_t, P)$ *and* $(\overline{\Omega}, \overline{\mathcal{F}}_t, \overline{P})$ *respectively. Then* Z *given by*

$$Z(t,w,\overline{w}) = X(t,w)\, Y(t,\overline{w})$$

is a martingale relative to

$$(\Omega \times \overline{\Omega}, \mathcal{F}_t \times \overline{\mathcal{F}}_t, P \times \overline{P}).$$

Proof. From the definition it is clear that for every $t > s$

$$\int_{A \times \overline{A}} Z(t,w,\overline{w})\, d(P \times \overline{P})\Big|_{\overline{\mathcal{F}_s \times \overline{\mathcal{F}}_s}} = \int_{A \times \overline{A}} Z(s,w,\overline{w})\, d(P \times \overline{P})$$

if $A \in \mathcal{F}_s$ and $\overline{A} \in \overline{\mathcal{F}}_s$. The general case follows easily.

As a corrollary to Lemma 2, we have

LEMMA 3. *Let* X *be a* d *-dimensional Itô process with parameters* b *and* a *relative to* $(\Omega, \mathcal{F}_t, P)$ *and let* Y *be a* $\overline{d}$ *-dimensional Itô process relative to* $(\overline{\Omega}, \overline{\mathcal{F}}_t, \overline{P})$ *relative to* $\overline{b}$ *and* $\overline{a}$. *Then* $Z(t,w,\overline{w}) = (X(t,w), Y(t,\overline{w}))$ *is a* $(d + d)$ *-dimensional Itô process with parameters* $B = (b, \overline{b})$, $A = \begin{bmatrix} a & 0 \\ 0 & \overline{a} \end{bmatrix}$ *relative to* $(\Omega \times {}^{-}, \mathcal{F}_t \times \overline{\mathcal{F}}_t, P \times \overline{P}$

LEMMA 4. *Let* X *be an Itô process relative to* $(\Omega, \mathcal{F}_t, P)$ *with parameters* 0 *and* a. *If* θ *is progressively measurable such that* $E(\int_0^t |\theta|_s^2 ds) < \infty, \forall\, t$ *and* $\theta a \theta^*$ *is bounded, then* $\int_0^t < \theta, dX > \in I[0, \theta a \theta^*]$.

Proof. Let θ_n be defined by

$$\theta_n^i = \begin{cases} \theta^i, & \text{if } |\theta| \leq n, \\ 0, & \text{otherwise} ; \end{cases}$$

Then $\int_0^t < \theta_n, dX > \in I[0, \theta_n\, a\, \theta_n^*]$. Therefore

$$X_n(t) = \exp\,(\lambda \int_0^t < \theta_n, dX > - \frac{\lambda^2}{2} \int_0^t < \theta_n, a\theta_n > ds$$

is a martingale converging pointwise to

$$X(t) = \exp\,(\lambda \int_0^t < \theta, dX > - \frac{\lambda^2}{2} \int_0^t < \theta, a\theta > ds).$$

To prove that $\int_0^t < \theta, dX >$ is an Itô process we have only to show that $X_n(t)$ is uniformly integrable. Without loss of generality we may assume that $\lambda = 1$. Let $[0,T]$ be given

$$\begin{aligned} E(X_n^2(t,w)) &= E(\exp[2 \int_0^t < \theta_n, dX > - \int_0^t < \theta_n, a\theta_n > ds]) \\ &= E(\exp[2 \int_0^t < \theta_n, dX > - 2 \int_0^t < \theta_n, a\theta_n > ds \\ &\qquad + \int_0^t < \theta_n, a\theta_n > ds]). \\ &\leq e^T \sup_{0 \leq t \leq T} < \theta_n, a\theta_n > . \end{aligned}$$

But $< \theta, a\theta^* >$ is bounded and therefore $< \theta_n, a\theta_n >$ is uniformly bounded in n. Therefore (X_n) are uniformly integrable . Thus $X(t,.)$ is a martingale.

Case (iii). Take $d = 1$, and assume that

$$\int_0^t a^{-1} \chi_{(a > 0)} ds < \infty, \forall\, t$$

with $a > 0$; let σ = +ve squareroot of a. Define $1/\sigma = 1/\sigma$ if $\sigma > 0$, and $1/\sigma = 0$ if $\sigma = 0$. Let

$$Y(t) = X(t) - x_0 - \int_0^t b(s,X(s))ds.$$

Denote by Z the one-dimensional Brownian motion on $(\Omega^*, \mathcal{F}_t^*, P^*)$ where $\Omega^* = C([0,\infty), R)$. Now

$$Y \in I[0,a(s,X(s,.))], Z \in I[0,1].$$

By Lemma 3,

$$(Y,Z) \in I\left[(0,0); \begin{pmatrix} a & 0 \\ 0 & 1 \end{pmatrix}\right].$$

If

$$\eta(t,w,w^*) = \int_0^t < (\frac{1}{\sigma(s,X(s,.))} \chi_{(\sigma > 0)}, \chi_{\sigma=0}, d(Y,Z)>$$

then Lemma 4 shows that

$$\eta \in I[0,1].$$

Therefore η is a one-dimensional Brownian motion on $\overline{\Omega} = (\Omega \times \Omega^*, \mathcal{F}_t \times \mathcal{F}_t^*, P \times P^*)$. Put

$$\overline{Y}(t,w,w^*) = Y(t,w) \text{ and } \overline{X}(t,w ,w^*) = X(t,w);$$

then

$$\int_0^t \sigma d\eta = \int_0^t \sigma \frac{1}{\sigma} \chi_{(\sigma > 0)} dY + \int_0^t \sigma \chi_{(\sigma = 0)} dZ$$

$$= \int_0^t \chi_{(\sigma > 0)} dY.$$

Since

$$E((\int_0^t \chi_{(\sigma = 0)} dY)^2) = E(\int_0^t \sigma^2 \chi_{(\sigma = 0)} ds) = 0 ,$$

it follows that

$$\int_0^t \sigma d\eta = \int_0^t dY = Y(t) = \overline{Y}(t,w,w^*).$$

Thus,

$$\overline{X}(t,w,w^*) = x_0 + \int_0^t \sigma(s,\overline{X}(s,w,w^*)d\eta +$$

$$+ \int_0^t b(s,\overline{X}(s,w,w^*)ds$$

with $\overline{X}(t,w,w^*) = X(t,w)$. Now

$$(P \times P^*) \overline{X}^{-1}(A) = (P \times P^*) (A \times \Omega^*) = P(A).$$

Therefore

$$(P \times P^*) \overline{X}^{-1} = P \quad \text{or} \quad Q\overline{X}^{-1} = P.$$

<u>Case (iv)</u>. (General Case). Define

$$Y(t,.) = X(t,.) - x_0 - \int_0^t b(s,X(s,.))ds.$$

Therefore $Y \in I[0,a(s,X(s,.))]$ relative to $(\Omega, \mathcal{F}_t, P)$. Let Z be the n-dimensional Brownian motion on $(\Omega^*, \mathcal{F}_t^*, P^*)$ where $\Omega^* = C([0,\infty);\ \mathbb{R}_n)$.

$$(Y,Z) \quad I[(0,0); \begin{pmatrix} a(s,X_s),0 \\ 0 \quad I \end{pmatrix}]$$

Let σ be a $d \times n$ matrix such that $\sigma\sigma^* = a$ on $[0,\infty) \times \mathbb{R}^d$. Let

$$\eta(t,w,w^*) = \int_0^t \sigma^{-1}(s,X(s,w))dY(s,w) + \int_0^t r_{N_\sigma}(s,Z(s,w^*))dZ(s,w^*)$$

$$= \int_0^t < (\sigma^{-1}(s,X(s,w)),\pi_{N_\sigma}(s,Z(s,w^*))),d\,(Y,Z) > .$$

Therefore η is an Ito process with parameters zero and

$$A = (\ ^{-1},\ _N) \begin{pmatrix} a & 0 \\ 0 & I \end{pmatrix} \begin{pmatrix} \sigma^{-1*} \\ \pi_{N_\sigma}^* \end{pmatrix}$$

$$= \sigma^{-1} a(\sigma^{-1})^* + \pi_{N_\sigma} \pi_{N_\sigma}^* .$$

$$= \pi_{N_\sigma} + \pi_{N_\sigma} \quad \text{(for any projection } PP^* = PP = P)$$

$$= I_{\mathbb{R}^n} .$$

Therefore η is n-dimensional Brownian motion on

$$(\bar\Omega, \bar{\mathcal{F}}_t, \bar P) = (\Omega \times \Omega^*, \mathcal{F}_t \times \mathcal{F}_t^*, P \times P^*).$$

$$\int_0^t \sigma(s,X(s,w))\, d\eta(s,w,w^*)$$

$$= \int_0^t \sigma(s,\bar X(s,w,w^*))d\eta(s,w,w^*), \text{where } \bar X(s,w,w^*) = X(s,w) ,$$

$$= \int_0^t \sigma\sigma^{-1}\, dY + \int_0^t \sigma\pi_{N_\sigma}\, dZ.$$

$$= \int_0^t \pi_{R_\sigma} \, dY, \text{ since } \sigma\sigma^{-1} = \pi_{R_\sigma} \text{ and } \sigma\pi_{N_\sigma} = 0,$$

$$= \int_0^t (1 - \pi_{R_\sigma}) \, dY.$$

<u>Claim</u>. $\int_0^t \pi_{R_\sigma} \, dY = 0.$

For

$$E[(\int_0^t \pi_{R_\sigma} \, dY)^2] = \int_0^t a\pi_{R_\sigma} \, ds = \int_0^t \sigma\sigma^* \, \pi_{R_\sigma} \, ds$$

$$= \int_0^t \sigma(0) \, ds = 0.$$

Therefore we obtain

$$\int_0^t \sigma(s,X(s,w))d\eta(s,w,w^*) = \int_0^t dY = Y(t) - Y(0) = Y(t)$$

putting $\overline{Y}(t,w,w^*) = Y(t,w)$, one gets

$$\overline{X}(t,w,w^*) = x_0 + \int_0^t \sigma(s,\overline{X}(s,w,w^*))d\eta(s,w,w^*) + \int_0^t b(s,\overline{X}(s,w,w^*))ds .$$

As in Case (iii) one shows easily that

$$(P \times P^*) \, \overline{X}^{-1} = P.$$

This completes the proof ot the theorem.

<u>COROLLARY</u>. *Let* $a: [0,\infty) \times \mathbb{R}^d \to S_d^+$, *and* $b: [0,\infty) \times \mathbb{R}^d \to \mathbb{R}^d$ *be bounded, progressively measurable functions. If for some choice of a Lipschitz function* $\sigma: [0,\infty) \times \mathbb{R}^d \to M_{d \times n}$, $\sigma\sigma^* = a$ *then the Itô process corresponding to* $[b,a)$ *is unique.*

To state the result precisely, let P_1 and P_2 be two probability measures on $C([0,\infty); \mathbb{R}^d)$ such that $X(t,w) = w(t)$ is an Itô process with parameters b and a. Then $P_1 = P_2$.

<u>Proof</u>. By the theorem, there exists a generalised n-dimensional Brownian motion β_i on $(\Omega_i, \Sigma_t^i, Q_i)$ and a map $\xi_i : \Omega_i \to \Omega$ satisfying (for $i = 1,2$)

$$_i(t,w) = x_0 + \int_0^t \sigma(s,\xi_i(s,w))d\beta_i(s,w) + \int_0^t b(s,\xi_i(s,w))ds.$$

and $P_i = Q_i\xi_i^{-1}$.

Now σ is Lipschitz so that ξ_i is unique but we know that the iterations converge to a solution. As the solution is unique the iterations converge to ξ_i. Each iteration is progressively measurable with respect to

$_t^i = \sigma\{\beta_i(s) : 0 \le s \le t\}$ so that ξ_i is also progressively measurable with respect to $\mathcal{F}_t^i$. Thus we can restate the result as follows: There exists $(\Omega_i, \mathcal{F}_t^i, Q_i)$ and a map $\xi_i : \Omega_i \to \Omega$ satisfying

$$\xi_i(t,w) = x_0 + \int_0^t \sigma(s,\xi_i(s,w))\, d\beta_i(s,w)$$

$$+ \int_0^t b(s,\xi_i(s,w))\, ds,$$

and $P_i = Q_i\xi_i^{-1}$.

$(\Omega_i, \mathcal{F}_t^i, Q_i, \beta_i)$ can be identified with the standard Brownian motion $(\Omega^*, \mathcal{F}_t^*, Q, \beta)$. Thus $P_1 = Q\xi^{-1} = P_2$, completing the proof.

27. ON LIPSCHITZ SQUARE ROOTS

LEMMA. *Let* $f\colon \mathbb{R} \to \mathbb{R}$ *be such that* $f(x) \geq 0$, $f(x) \in C^2$ *and* $|f''(x)| \leq A$ *on* $(-\infty,\infty)$; *then*

$$|f'(x)| \leq \sqrt{f(x)}\,\sqrt{2A}.$$

Proof. $0 \leq f(y) = f(x) + (y-x)f'(x) + \frac{(y-x)^2}{2} f''(\xi)$

$$\leq f(x) + Zf'(x) + \frac{Z^2}{2} f''(\xi)$$

where $Z = y - x$, or $f(y) \leq f(x) + Zf'(x) + \frac{AZ^2}{2}$. Therefore

$$\frac{AZ^2}{2} + Zf'(x) + f(x) \geq 0,\ \forall\, Z \in \mathbb{R}$$

$$|f'(x)|^2 \leq 2A\, f(x).$$

So

$$|f'(x)| \leq \sqrt{2A}\ \ f(x).$$

NOTE. If we take $f(x) = x^2$, we note that the constant is the best possible.

COROLLARY. *If* $f \geq 0$, $|f''| \leq A$, *then*

$$|\sqrt{(f(x_1))} - \sqrt{(f(x_2))} \leq \sqrt{(A/2)}\,|x_1 - x_2|.$$

Proof. Let $\varepsilon > 0$, then $\sqrt{(f(x)+\varepsilon)}$ is a smooth function.

$$(\sqrt{(f(x)+\varepsilon)})' = \frac{f'(x)}{2\sqrt{(f(x)+\varepsilon)}} = \frac{(f(x)+\varepsilon)'}{2\sqrt{(f(x)+\varepsilon)}}\,.$$

Therefore

$$|(\sqrt{(f(x)+\varepsilon)})'| \leq \sqrt{(2A/2)} \leq \sqrt{(A/2)}\ ,$$

or

$$|\sqrt{(f(x_1)+\varepsilon)} - \sqrt{(f(x_2)+\varepsilon)}| \leq \sqrt{(A/2)}\ |x_1 - x_2|.$$

Let $\varepsilon \to 0$ to get the result.

We now consider the general case and give conditions on the matrix a so that σ defined by $\sigma\sigma^* = a$ is Lipschitz.

<u>THEOREM</u>. *Let* $a: \mathbb{R}^n \to S_d^+$ *be continuous and bounded* C^2 *-function such that the second derivative is uniformly bounded, i.e.* $\|D_s D_r a_{ij}\| \leq M$, *where* M *is independent of* i,j,r,s; ($D_r \equiv \frac{d}{dx_r}$). *If* $\sigma: \mathbb{R}^n \to S_d^+$ *is the unique positive square root of* a, *then*

$$\|\sigma(x_1) - \sigma(x_2)\| \leq A|x_1 - x_2|, \forall\ x_1, x_2, A = A(M,d).$$

<u>Proof</u>. <u>Step 1</u>. Let $A \in S_d^+$ be strictly positive such that $\|I - A\| < 1$. Then

$$\sqrt{A} = \sqrt{(I - (I - A))}$$

$$= \sum_{r=0}^{\infty} \frac{C_r}{r!} (I - A)^r,$$

so that on the set $\{A: \|I - A\| < 1\}$ the map $A \to \sqrt{A}$ is C^∞ (in fact analytic).

Now assume that A is any positive definite matrix. Let $\lambda_1,..,\lambda_n$ be the eigen values so that $\lambda_j > 0$, $j = 1,2..n$. Therefore $I - \varepsilon A$ is aymmetric with eigen values $1 - \varepsilon\lambda_j$. By **choosing** ε sufficiently small we can make

$$\|I - \varepsilon A\| = \max\{1 - \varepsilon\lambda_1, .., 1 - \varepsilon\lambda_n\} < 1.$$

Fixing such an ε we observe that

$$\sqrt{A} = \frac{1}{\sqrt{\varepsilon}} \sqrt{(\varepsilon A)} = \frac{1}{\sqrt{\varepsilon}} \sqrt{(I - (I - \varepsilon A))}.$$

So the map $A \to \sqrt{A}$ is smooth on the set of symmetric positive definite matrices.

Step 2. Let $n = 1$, $\sigma(t_0) = \sqrt{(a(t_0)}$ where $a(t_0)$ is positive definite. Assume $a(t_0)$ to be diagonal so that $\sigma(t_0)$ is also diagonal.

$$\sum_j \sigma_{ij}(t)\, \sigma_{jk}(t) = a_{ik}(t).$$

Differentiating with respect to t at $t = t_0$ we get

$$\sum_j \sigma_{ij}(t_0)\, \sigma_{jk}'(t_0) + \sum_j \sigma_{ij}'(t_0)\sigma_{jk}(t_0) = a_{ik}'(t_0)$$

or

$$\sqrt{a_{ii}}(t_0)\sigma_{ik}'(t_0) + \sqrt{a_{kk}}(t_0)\sigma_{ik}'(t_0) = a_{ik}'(t_0)$$

or

$$\sigma_{ik}'(t_0) = \frac{a_{ik}'(t_0)}{\sqrt{(a_{ii}(t_0))} + \sqrt{(a_{kk}(t_0))}}\,.$$

Since the second derivatives are bounded by $4M$ and $a_{ii} - 2a_{ij} + a_{jj} \geq 0$, we get

$$|a_{ii}'(t) + 2a_{ij}'(t) + a_{jj}'(t)| \leq \sqrt{(8M)}\sqrt{(a_{ii}(t) + 2a_{ij}(t) + a_{jj}(t))}$$

$$\leq \sqrt{(8M)}\sqrt{2}\,\sqrt{(a_{ii} + a_{jj})}(t)$$

or

(1) $$|a_{ii}'(t) + 2a_{ij}'(t) + a_{jj}'(t)| \leq 4\sqrt{M}\,(\sqrt{a_{ii}} + \sqrt{a_{jj}}).$$

Since a is non-negative definite,

$$|a''_{ii}(t)| \leq \sqrt{(2M)}\ \sqrt{(a_{ii}(t))},\ \forall\ i.$$

substituting this in (1), we get

$$|a'_{ij}(t)| \leq 4\sqrt{M}\ (\sqrt{a_{ii}} + \sqrt{a_{jj}}),$$

and hence

$$|\sigma'_{ij}(t_0)| \leq 4\sqrt{M}.$$

Step 3. Let $a(t_0)$ be positive definite and σ its positive definite square root. There exists a constant unitary matrix α such that $\alpha a(t_0)\,\alpha^{-1} = b(t_0)$ is a diagonal positive definite matrix. Let $\lambda(t_0)$ be the positive square root of $b(t_0)$ so that

$$\lambda(t_0) = \alpha\sigma(t_0)\,\alpha^{-1}.$$

Therefore $\sigma'(t_0) = (\alpha^{-1}\lambda'\alpha)(t_0)$ where $(\sigma'(t_0))_{ij} = \sigma'_{ij}(t_0)$ and $a''(t_0) = (\alpha^{-1}b''\alpha)(t_0)$. Since α is unitary,

$$\|\lambda\| = \|\sigma\|,\ \|a''\| = \|b''\|,\|\lambda'\| = \|\sigma'\|.$$

By hypothesis, $\|b''\| = \|a''\| \leq C(d).M$. Therefore $\|\lambda'\| \leq 4\sqrt{(MC(d))}$, i.e. $\|\sigma'\| \leq 4\sqrt{(MC(d))}$.

Thus $\|\sigma(t_1) - \sigma(t_2)\| \leq |t_1 - t_2|C(M,d)$.

Step 4. Let $a\colon \mathbb{R} \to S_d^+$ and σ be the unique non-negative definite square root of a. For each $\varepsilon > 0$ let $a_\varepsilon = a + \varepsilon I, \sigma_\varepsilon$ = unique positive square root of a_ε. Then by step 3,

$$\|\sigma_\varepsilon(t_1) - \sigma_\varepsilon(t_2)\| \leq C(M,d)|t_1 - t_2|.$$

If a is diagonal then it is obvious that $\sigma_\varepsilon \to \sigma$ as $\varepsilon \to 0$. In the general case reduce **a** to the diagonal form and conclude that $\sigma_\varepsilon \to \sigma$.

Thus

$$\|\sigma(t_1) - \sigma(t_2)\| \le C(M,d) \, |t_1 - t_2| .$$

Step 5. Let $a: \mathbb{R}^n \to S_d^+$ and $\sigma^2 = a$. with $\|D_r D_s a_{ij}\| \le M$, $\forall$ x, i,j; r,s $x \in \mathbb{R}^n$. Choose $x_1, x_2 \in \mathbb{R}^n$. Let $x_1 = y_1, y_2 \ldots, y_{n+1} = x_2$ be $(n + 1)$ points such that y_i and y_{i+1} differ almost in one coordinate. By Step 4, we have

$$(*) \qquad \|\sigma(y_i) - \sigma(y_{i+1})\| \le C|y_i - y_{i+1}| .$$

The result follows easily from the fact that

$$\|x\|_1 = \sum_{i=1}^{n} |x_i| \quad \text{and} \quad \|x\|_2 = (x_1 + .. + x_n)^{1/2}$$

are equivalent norms.

This completes the proof of the theorem.

28. RANDOM TIME CHANGES

LET

$$L = \frac{1}{2} \sum_{i,j} a_{ij} \frac{\partial^2}{\partial x_i \partial x_j} + \sum_j b_j \frac{\partial}{\partial x_j}$$

with $a: \mathbb{R}^d \to S_d^+$ and $b: \mathbb{R}^d \to \mathbb{R}^d$ bounded measurable functions. Let $X(t,.)$, given by $X(t,w) = w(t)$ for (t,w) in $[0,\infty) \times C([0,\infty): \mathbb{R}^d$ be an Itô process corresponding to $(\Omega, \mathcal{F}_t, Q)$ with parameters b and a where $\Omega = C([0,\infty); \mathbb{R}^d)$. For every constant $c > 0$ define

$$L_c \equiv c[\frac{1}{2} \sum_{i,j} a_{ij} \frac{\partial^2}{\partial x_i \partial x_j} + \sum_j b_j \frac{\partial}{\partial x_j}].$$

Define Q_c by $Q_c = PT_c^{-1}$ where $(T_c w)(t) = w(\,ct)$. Then one can show that X is an Itô process corresponding to $(\Omega, \mathcal{F}_t, Q_c)$ with parameters cb and ca [Note: We have done this in the case where $a_{ij} = \delta_{ij}$].

Consider the equation

$$\frac{\partial u}{\partial t} = L_c u \quad \text{with} \quad u(0,x) = f(x).$$

This can be written as $\frac{\partial u}{\partial \tau} = Lu$ with $u(0,x) = f(x)$ when $\tau = ct$. Thus changing time in the differential equation is equivalent to stretching time in probablistic language.

So far we have assumed that c is a constant. Now we shall allow c to be a function of x.

Let $\phi: \mathbb{R}^d \to \mathbb{R}$ be any bounded measurable function such that

$$0 < C_1 \le \phi(x) < C_2 < \infty, \ \forall\, x \in \mathbb{R}^d$$

and suitable constants C_1 and C_2. If

$$L \equiv \left[\frac{1}{2} \sum a_{ij} \frac{\partial^2}{\partial x_i \partial x_j} + \sum b_j \frac{\partial}{\partial x_j}\right]$$

we define

$$L_\phi = \phi L \equiv \phi\left[\frac{1}{2} \sum a_{ij} \frac{\partial^2}{\partial x_i \partial x_j} + \sum b_j \frac{\partial}{\partial x_j}\right].$$

In this case we can say that the manner which time changes depends on the position of the particle.

Define $T_\phi : \Omega \to \Omega$ by

$$(T_\phi w)(t) = w(\tau_t(w))$$

where $\tau_t(w)$ is the solution of the equation

$$\int_0^{\tau_t} \frac{ds}{\phi(w(s))} = t.$$

As $C_1 \leq \phi \leq C_2$ it is clear that $\tau_t \frac{1}{C_1} \leq t \leq \tau_t \frac{1}{C_2}$. When $\phi \equiv c$ a constant, then $\tau_t = ct$ and T_ϕ coincides with T_c.

As

$$0 < C_1 \leq \phi \leq C_2 < \infty, \int_0^\lambda \frac{1}{\phi(w(s))} ds$$

is continuous and increases strictly from 0 to ∞ as λ increases, so that τ_t exists, is unique, and is a continuous function of t for each fixed w.

Some properties of T_ϕ.

(i) If 1 is the constant function taking the value 1 then it is clear that T_1 = identity.

(ii) Let ϕ and ψ be two measurable functions such that $0 < a \leq \phi(x), \psi(x) \leq b < \infty$, $\forall\, x \in \mathbb{R}^d$. Then $T_\phi 0\, T = T\phi\psi = T_\psi 0\, T_\phi$.

Proof. Fix w. Let τ_t be given by

$$\int_0^{\tau_t} \frac{1}{\phi(w(s))}\, ds = t.$$

Let $w^*(t) = w(\tau_t)$ and let σ_t be given by

$$\int_0^{\sigma_t} \frac{1}{\psi(w^*(s))}\, ds = t.$$

Let $w^{**}(t) = w^*(\sigma_t) = w(\tau_{\sigma_t})$. Therefore

$$((T_\psi\, 0\, T_\phi)w)(t) = (T_\psi w^*)(t) = w^*(\sigma_t)$$

$$= w^{**}(t) = w(\tau_{\sigma_t}).$$

Hence to prove the property (ii) we need only show that

$$\int_0^{\tau_{\sigma_t}} \frac{1}{\phi(w(s))} \frac{1}{\psi(w(s))}\, ds = t.$$

Since

$$\int_0^{\tau_t} \frac{1}{\phi(w(s))}\, ds = t, \quad \frac{dt}{d\tau_t} = \frac{1}{\phi(w(\tau_t))}$$

and

$$\frac{dt}{d\sigma_t} = \frac{1}{\psi(w^*(\sigma_t))} = \frac{1}{\psi(w(\tau_{\sigma_t})}.$$

Therefore

$$\frac{d\tau_{\sigma_t}}{dt} = \frac{d\tau_{\sigma_t}}{d\sigma_t} \cdot \frac{d\sigma_t}{dt} = \phi(w(\tau_{\sigma_t}))\, \psi(w^*(\sigma_t))$$

$$= \phi(w(\tau_{\sigma_t}))\, \psi(w(\tau_{\sigma_t}))$$

$$= (\phi\psi)\,(w(\tau_{\sigma_t})).$$

Thus

$$\int_0^{\tau_{\sigma_t}} \frac{1}{(\phi\psi)(w(s))}\, ds = t.$$

This completes the proof.

(iii) From (i) and (ii) it is clear that $T_\phi^{-1} = T_{\phi-1}$ where $\phi^{-1} = \frac{1}{\phi}$.

(iv) (τ_t) is a stopping time relative to ${}_t$. i.e.

$$\left\{w: \int_0^\lambda \frac{1}{\phi(w(s))}\, ds \geq t\right\} \in {}_\lambda \quad \text{for each } \lambda \geq 0.$$

(v) $$T_\phi(w)(t) = w(\tau_t w) = X_{\tau_t}(w).$$

Thus T_ϕ is $(\mathcal{F}_t - \mathcal{F}_{\tau_t})$-measurable, i.e. $T_\phi^{-1}(\mathcal{F}_t) \subset \mathcal{F}_{\tau_t}$.

Since $X(t)$ is an Itô process, with parameters b, a, $\forall f \in C_0^\infty(\mathbb{R}^d)$, $f(X(t)) - \int_0^t (Lf)(X(s))ds$ is a martingale relative to $(\Omega, \mathcal{F}_t, P)$. By the optional sampling theorem

$$f(X_{\tau_t}) - \int_0^{\tau_t} (Lf)\,(X(s))ds$$

is a martingale relative to $(\Omega, \mathcal{F}_{\tau_t}, P)$, i.e.

$$f(X_{\tau_t}) - \int_0^t (Lf)\,(X(\tau_s))\, d\tau_s$$

is a martingale relative to $(\Omega, \mathcal{F}_{\tau_t}, P)$. But $\frac{d\tau_s}{dt} = \phi$. Therefore

$$f(X(\tau_t)) - \int_0^t (Lf)(X_{\tau_s})\,\phi(X_{\tau_s})\,ds$$

is a martingale.

Put $Y(t) = X_{\tau_t}$ and appeal to the definition of L_ϕ to conclude that

$$f(Y(t)) - \int_0^t (L_\phi f)(Y(s))ds$$

is a martingale. $Y(t,w) = X_{\tau_t}(w) = (T_\phi w)(t)$. Let $\overline{\mathcal{F}}_t = \sigma\{Y(s): 0 \leq s \leq t\}$. Then $\overline{\mathcal{F}}_t = T_\phi^{-1}(\mathcal{F}_t) \subset \mathcal{F}_{\tau_t}$. Thus

$$f(Y(t)) - \int_0^t (L_\phi f)\,(Y(s))ds$$

is a martingale relative to $(\Omega, \overline{\mathcal{F}}_t, P)$. Define $Q = PT_\phi^{-1}$ so that

$$f(X(t)) - \int_0^t (L_\phi f)\,(X(s))ds$$

is an $(\Omega, \mathcal{F}_t, Q)$ -martingale, i.e. Q is an Itô process that corresponds to the operator ϕL. Or, PT_ϕ^{-1} is an Itô process that corresponds to the operator ϕL.

We have now proved the following theorem.

<u>THEOREM</u>. *Let* $\Omega = C([0,\infty); \mathbb{R}^d)$; $X(t,w) = w(t)$;

$$L = \frac{1}{2}\sum_{i,j} a_{ij}\frac{\partial^2}{\partial x_i \partial x_j} + \sum b_j \frac{\partial}{\partial x_j}.$$

Suppose that $X(t)$ *is an Itô process relative to* $(\Omega, {}_t, P)$ *that corresponds to the operator* L. *Let* $0 \leq C_1 \leq \phi \leq C_2$ *where*

$\phi: \mathbb{R}^d \to \mathbb{R}$ *is measurable. If* $Q = PT_\phi^{-1}$, *then* $X(t)$ *is an Itô process relative to* $(\Omega, \mathcal{F}_t, Q)$ *that corresponds to the operator* ϕL.

As $0 < C_1 \leq \phi \leq C_2$, we get $0 < 1/C_2 \leq 1/\phi < 1/C_1$ with $T_{\phi^{-1}} \circ T_\phi = I$. We have thus an obvious corollary.

<u>COROLLARY</u>. *There exists a probability measure* P *on* Ω *such that* X *is an Itô process relative to* $(\Omega, \mathcal{F}_t, P)$ *that corresponds to the operator* L *if and only if there exists a probability measure* Q *on* Ω *such that* X *is an Ito process relative to* $(\Omega, \mathcal{F}_t, Q)$ *that corresponds to the operator* ϕL.

<u>REMARK</u>. If $C_2 \geq \phi \geq C_1 > 0$ then we have shown that existence and uniqueness of an Itô process for the operator L guarantees existence and uniqueness of the Itô process for the operator ϕL. The solution is no longer unique if we relax the strict positivity on C_1 as is illustrated by the following example.

Let $\phi \equiv a(x) = |x|^\alpha \wedge 1$ where $0 < \alpha < 1$ and let $L = \frac{1}{2} a \frac{\partial^2}{\partial x}$. Define δ_0 on $\{C([0,\infty); \mathbb{R})\}$ by

$$\delta_0(A) = \begin{cases} 1, & \text{if } \theta \in A, \forall A \in \quad , \\ 0, & \text{if } \theta \notin A, \end{cases}$$

where θ is the zero function on $[0,\infty)$.

<u>Claim</u>. δ_0 is an Itô process with parameters 0 and a. For this it is enough to show that, $\forall f \in C_0^\infty(\mathbb{R})$

$$f(X(t)) - \int_0^t (Lf)(X(s))ds$$

is a martingale, using $a(0) = 0$. it follows easily that

$$\int_A \int_0^t (Lf)(X(\sigma))d\sigma d\delta_0 = 0$$

$\forall$ Borel set A of $C([0,\infty); \mathbb{R})$ and $\int_A f(X(t))d\delta_0 = 0$ if $\theta \notin A$ and $\int_A f(X(t))d\delta_0 = f(0)$ if $\theta \in A$, and th is is true $\forall$ t, showing that $X(t,w) = w(t)$ is an Itô process relative to δ_0 corresponding to the operator L.

Next we shall define T_a (as in the theorem); we note that T_a cannot be defined everywhere (for example $T_a(\theta)$ is not defined). However T_a is defined a.e. P where $P = P_0$ is the Brownian motion.

$$E^P\left(\int_0^t \frac{1}{|X(s)|^\alpha} ds\right) = \int_0^t \int_0^\infty \frac{1}{y^\alpha} \frac{1}{\sqrt{(2\pi s)}} e^{-\frac{y}{2s}} dy\, ds < \infty$$

since $0 < \alpha < 1$. Thus by Fubini's theorem,

$$\int_0^t \frac{1}{|w(s)|^\alpha} ds < \infty \text{ a.e.}$$

Taking $t = 1,2,3..$, there exists a set Ω^* such that $P(\Omega^*) = 1$ and

$$\int_0^t \frac{1}{|w(s)|^\alpha} ds < \infty, \forall t, \forall w \in \Omega^*$$

Observe that

$$\int_0^t \frac{1}{|w(s)|^\alpha} ds < \infty$$

implies that

$$\int_0^t \frac{1}{|w(s)|^{\alpha} \wedge 1}\, ds < \infty,$$

for

$$\int_0^t \frac{ds}{|w(s)|^{\alpha} \wedge 1} =$$

$$= \int_{[0,t] \cap \{|w(s)|^{\alpha} > 1\}} \frac{ds}{|w(s)|^{\alpha} \wedge 1} + \int_{\{|w(s)|^{\alpha} \leq 1\} \cap [0,t]} \frac{ds}{|w(s)|^{\alpha}} \cdots\cdots$$

$$\leq m\, \{(|w(s)|^{\alpha} > 1) \cap [0,t]\} + \int_0^t \frac{1}{|w(s)|^{\alpha}} ds < \infty$$

(m = Lebesgue measure)

Thus T_a is defined on the whole of Ω^*. Using the same argument as in the theorem, it can now be proved that X is an Itô process relative to Q corresponding to the operator L. Finally, we show that $Q\{\theta\} = 0$. $Q\{\theta\} = PT_a^{-1}\{\theta\}$. Now: $T_a^{-1}\{\theta\}$ = empty. For, let $w \in T_a^{-1}\{\theta\}$. Then $w(\tau_t) = 0$, $\forall\, t$, $w \in \Omega^*$. Since $|\tau_t - \tau_s| \leq |t - s|$, one finds that τ_t is a continuous function of t. Further $\tau_1 > 0$, and $w = 0$ on $[0,\tau_1]$ gives

$$\int_0^{\tau_1} \frac{1}{|w(s)|^{\alpha} \wedge 1}\, ds = \infty.$$

This is false unless $T_a^{-1}\{\theta\}$ = empty. Thus $Q\{\theta\} = 0$ and Q is different from δ_0.

29. CAMERON - MARTIN - GIRSANOV FORMULA

LET US REVIEW what we did in Bronian motion with drift.

Let $(\Omega, \mathcal{F}_t, P)$ be a d-dimensional Brownian motion with

$$P\{w: w(0) = x\} = 1.$$

Let $b: \mathbb{R}^d \to \mathbb{R}^d$ be a bounded measurable function and define

$$Z(t) = \exp\left(\int_0^t \langle b, dx \rangle - \frac{1}{2}\int_0^t |b|^2 ds\right).$$

Then we saw that $Z(t,.)$ is an $(\Omega, \mathcal{F}_t, P)$-martingale. We then had a probability measure Q given by the formula

$$\left.\frac{dQ}{dP}\right|_{\mathcal{F}_t} = Z(t,.).$$

We leave it as an exercise to check that in effect X is an Itô process relative to Q with parameters b and I. In other words we had made a transition from the operator $\Delta/2$ to $\Delta/2 + b.\nabla$. We now see whether such a relation also exists for the more general operator L.

<u>THEOREM</u>. *Let* $a: \mathbb{R}^d \to S_d^+$ *be bounded and measurable such that* $a \geq CI$ *for some* $C > 0$. *Let* $b: \mathbb{R}^d \to \mathbb{R}^d$ *be bounded,* $\Omega = C([0,\infty); \mathbb{R}^d)$, $X(t,w) = w(t)$, P *any probability measure on* Ω *such that* X *is an Itô process relative to* $(\Omega, \mathcal{F}_t, P)$ *with parameters* $[0,a]$. *Define* Q^t *on* $\mathcal{F}_t$ *by the rule*

$$\frac{dQ^t}{dP}\Big| \mathcal{F}_t = Z(t,.) = \exp[\int_0^t < a^{-1}b, dX > - \frac{1}{2}\int_0^t < b, a^{-1}b > ds].$$

Then

(i) $\{Q^t\}$ $t \geq 0$ *is a consistent family.*

(ii) *there exists a measure* Q *on* $\sigma(\bigcup \mathcal{F}_t)$:

$$Q\big|\mathcal{F}_t = Q^t.$$

(iii) X(t) *is an Itô process relative to* $(\Omega, \mathcal{F}_t, Q)$ *with parameters* [b,a], *i.e. it corresponds to the operator*

$$\frac{1}{2}\sum_{i,j} a_{ij}\frac{\partial^2}{\partial x_i \partial x_j} + \sum_j b_j \frac{\partial}{\partial x_j}.$$

<u>Proof.</u> (i) Let $A(t) = \int_0^t < a^{-1}b, dX >$. Then $A \in I[0, < b, a^{-1}b >]$. Therefore Z(t) is a martingale relative to $(\Omega, \mathcal{F}_t, P)$ hence $\{Q^t\}_{t \geq 0}$ is a consistent family.

(ii) Proof as in the case of Brownian motion.

(iii) We have to show that

$$\exp[< \theta, X(t,.) > - < \theta, \int_0^t b\,ds > - \frac{1}{2}\int_0^t < \theta, a\theta > ds]$$

is a martingale relative to $(\Omega, \mathcal{F}_t, Q)$.

Now for any function θ which is progressively measurable and bounded

$$\exp[\int_0^t < \theta, dX > - \frac{1}{2}\int_0^t < \theta, a\theta > ds]$$

is an $(\Omega, \mathcal{F}_t, P)$-martingale. Replace θ by

$\theta(w) = \theta + (a^{-1}b)(X(s,w))$, where θ now is a constant vector. Then

$$\exp\Big[\int_0^t < \theta + a^{-1}b, dX > - \frac{1}{2}\int_0^t < \theta + a^{-1}b, a\theta > ds$$

is an $(\Omega, \mathcal{F}_t, P)$-martingale, i.e.

$$\exp\Big[< \theta, X(t) > - \frac{1}{2}\int_0^t < \theta, a\theta > ds - \frac{1}{2}\int_0^t < a^{-1}b, a\theta > - \frac{1}{2}\int_0^t < \theta, b >\Big]$$

is an $(\Omega, \mathcal{F}_t, Q)$-martingale, and

$$\begin{aligned} < a^{-1}b, a\theta > &= < a^* a^{-1}b, \theta > \\ &= < aa^{-1}b, \theta > \qquad (\text{since } a = a^*) \\ &= < b, \theta >. \end{aligned}$$

Thus

$$\exp\Big[< \theta, X(t) > - \frac{1}{2}\int_0^t < \theta, a\theta > ds - \int_0^t < \theta, b > ds\Big]$$

is an $(\Omega, \mathcal{F}_t, Q)$-martingale, i.e. X is an Itô process relative to $(\Omega, \mathcal{F}_t, Q)$ with parameters $[b,a]$. This proves the theorem.

We now prove the converse part.

<u>THEOREM</u>. *Let*

$$L_1 = \frac{1}{2}\sum_{i,j} a_{ij}\frac{\partial^2}{\partial x_i \partial x_j}$$

and

$$L_2 \equiv \frac{1}{2}\sum_{i,j} a_{ij}\frac{\partial^2}{\partial x_i \partial x_j} + \sum_j b_j \frac{\partial}{\partial x_j},$$

where $a: \mathbb{R}^d \to S_d^+$ *is bounded measurable such that* $a \geq CI$ *for some* $C > 0$; $b: \mathbb{R}^d \to \mathbb{R}^d$ *is bounded and measurable.. Let* $\Omega = C([0,\infty); \mathbb{R}^d)$ *with* $\mathcal{F}_t$ *as usual. Let* θ *be a probability measure on* $\sigma(\bigcup \mathcal{F}_t)$

and X *a progressively measurable function such that* X *is an Itô process relative to* $(\Omega, \mathcal{F}_t, Q)$ *with parameters* $[b,a]$ *i.e.* X *corresponds to the operator* L_2. *Let*

$$Z(t) = \exp \left[- \int_0^t < a^{-1}b, dX > + \frac{1}{2} \int_0^t < b, a^{-1}b > ds\right].$$

Then

(i) $Z(t)$ *is an* $(\Omega, \mathcal{F}_t, Q)$ *-martingale.*

(ii) *If* P^t *is defined on* $\mathcal{F}_t$ *by*

$$\frac{dP^t}{dQ}\bigg|\mathcal{F}_t = Z(t),$$

Then there exists a probability measure P *on* $\sigma(\bigcup \mathcal{F}_t)$ *such that*

$$P\big|\mathcal{F}_t = P^t$$

(iii) X *is an Itô process relative to* $(\Omega, \mathcal{F}_t, P)$ *corresponding to parameters* $[0,a]$, *i.e.* X *corresponds to the operator* L_1.

<u>Proof</u>. (i) Let

$$A(t) = \int_0^t < -a^{-1}b, dX >.$$

Then $A(t)$ is an Itô process with parameters $[< -a^{-1}b, b >, < a^{-1}b, b >]$.

Thus

$$\exp \left[A(t) - \int_0^t < -a^{-1}b, b > ds - \frac{1}{2} \int_0^t < a^{-1}b, b > ds\right]$$

is an $(\Omega, \mathcal{F}_t, Q)$ -martingale, i.e. $Z(t)$ is an $(\Omega, \mathcal{F}_t, Q)$ martingale.

(ii) By (i), P^t is a consistent family. The proof that there exists a probability measure P is same as before.

Since X is an Itô process relative to Q with parameters b and a,

$$\exp\left[\int_0^t <\theta, dX> - \int_0^t <\theta, b> ds - \frac{1}{2}\int_0^t <\theta, a\theta> ds\right]$$

is a martingale relative to Q for every bounded measurable θ. Replace θ by $\theta(w) = \theta - (a^{-1}b)(X(s,w))$ where θ now is a constant vector to get

$$\exp\left[<\theta, X(t)> - \int_0^t < a^{-1}b, dX> - \int_0^t <\theta, b> + \int_0^t < a^{-1}b, b> ds - \right.$$

$$\left. - \frac{1}{2}\int_0^t < \theta - a^{-1}b, a\theta - b> ds\right]$$

is an $(\Omega, \mathcal{F}_t, Q)$ martingale, i.e.

$$\exp\left[<\theta, X> - \int_0^t < a^{-1}b, dX> - \int_0^t <\theta, b> ds + \int_0^t < a^{-1}b, b> ds - \right.$$

$$- \frac{1}{2}\int_0^t <\theta, a\theta> ds - \frac{1}{2}\int_0^t < a^{-1}b, b> ds + \frac{1}{2}\int_0^t <\theta, b> ds +$$

$$\left. + \int_0^t < a^{-1}b, a\theta> ds\right]$$

is an $(\Omega, \mathcal{F}_t, Q)$ martingale. Let $\theta \in \mathbb{R}^d$, so that

$$\exp\left[<\theta, X> - \frac{1}{2}\int_0^t <\theta, b> ds - \frac{1}{2}\int_0^t <\theta, a\theta> ds + \frac{1}{2}\int_0^t < a^{-1}b, a\theta> ds\right] Z(t)$$

is an $(\Omega, \mathcal{F}_t, Q)$ -martingale and

$$< a^{-1}b, a> = < b, \theta> \quad (\text{since } a = a^*).$$

Therefore

$$\exp \left[< \theta, X > - \frac{1}{2} \int_0^t < \theta, a\theta > ds\right] Z(t)$$

is an $(\Omega, \mathcal{F}_t, Q)$ martingale.

Using the fact that $\left.\frac{dP}{dQ}\right| \mathcal{F}_t = Z(t)$, we conclude that

$$\exp \left[< \theta, X > - \frac{1}{2} \int_0^t < \theta, a\theta > ds\right]$$

is a martingale relative to $(\Omega, \mathcal{F}_t, P)$, i.e. $X \in I[0,a]$ relative to $(\Omega, \mathcal{F}_t, P)$. This proves the theorem.

SUMMARY. We have the following situation

$$L_1, \Omega, \mathcal{F}_t, \quad \Omega = C([0,\infty); \mathbb{R}^d), L_2, \Omega, \mathcal{F}_t.$$

P a probability measure such that X is an Itô Process relative to P corresponding to the operator L_1.	$\Rightarrow$	X is an Itô process relative to a probability measure Q corresponding to L_2. Q is given by $\left.\frac{dQ}{dP}\right| \mathcal{F}_t = Z(t,.)$
X is an Itô process relative to P corresponding to L_1 where $\left.\frac{dP}{dQ}\right| \mathcal{F}_t = \frac{1}{Z(t,.)}$	$\Leftarrow$	X is an Itô process relative to Q corresponding to L_2.

Thus existence and uniqueness for any sy stem guarantees the existence and uniqueness for the other system.

APPLICATION. (Exercise).

Take $d = 1$, $a\colon \mathbb{R} \to \mathbb{R}$ bounded and measurable with $0 < C_1 \leq a < C_2 < \infty$. Let $L = \frac{a}{2} \frac{\partial^2}{\partial x^2} + b \frac{\partial}{\partial x}$. Show that there exists a unique probability measure P on $\Omega = C([0,\infty); \mathbb{R})$ such that $X(t)$ is Itô relative to P corresponding to L.

$(X(t,w) \equiv w(t))$ for any given starting point.

30. BEHAVIOUR OF DIFFUSIONS FOR LARGE TIMES

LET $L_2 = \Delta/2 + b.\nabla$ WITH $b: \mathbb{R}^d \to \mathbb{R}^d$ measurable and bounded on each compact set. We assume that there is no explosion. If P_x is the d-dimensional Brownian measure on $\Omega = C([0,\infty); \mathbb{R}^d)$ we know that there exists a probability measure Q_x on Ω such that

$$\left.\frac{dQ_x}{dP_x}\right|_t = \exp\left[\int_0^t < b, dX > - \frac{1}{2}\int_0^t |b|^2 ds\right]$$

Let K be any compact set in $\mathbb{R}^d$ with non-empty interior. We are interested in finding out how often the trajectories visit K and whether this 'frequency' depends on the starting point of the trajectory and the compact set K.

THEOREM. *Let* K *be any compact set in* $\mathbb{R}^d$ *having a non-empty interior. Let*

$$E_\infty^K = \{w: w \text{ revisits } K \text{ for arbitrarily large times}\}$$
$$= \{w: \text{there exists a sequence } t_1 < t_2 < . < \infty \text{ with } t_n \to \infty \text{ such that } w(t_n) \in K\}$$

Then,

either $Q_x(E_\infty^K) = 0, \forall x,$ *and* $\forall K,$

or $Q_x(E_\infty^K) = 1, \forall x,$ *and* $\forall K.$

REMARK. 1. In the first case $\lim_{t \to +\infty} |X(t)| = +\infty$ a.e. $Q_x, \forall x,$ i.e. almost all trajectories stay within K only for a short period.

These trajectories are called *transient*. In the second case almost all trajectories visit K for arbitrary large times. Such trajectories are called *recurrent*.

2. If $b = 0$ then $Q_x = P_x$. For the case $d = 1$ or $d = 2$ we know that the trajectories are recurrent. If $d \geq 3$ the trajectories are transient.

Proof. Step 1. We introduce the following sets.

$$E_0^K = \{w: X(t,w) \in K \text{ for some } t \geq 0\},$$

$$E_{t_0}^K = \{w: X(t,w) \in K \text{ for some } t \geq t_0\}, \quad 0 \leq t_0 < \infty.$$

Then clearly

$$E_\infty^K = \bigcap_{n=1}^{\infty} E_n^K = \bigcap_{t_0 \geq 0} E_{t_0}^K .$$

Let

$$\psi(x) = Q_x(E_\infty^K), \quad F = \chi_{E_\infty^K} .$$

$$E^{Q_x}(F|\mathcal{F}_t) = E^{Q_x}(\chi_{E_\infty^K}|\mathcal{F}_t) = Q_{X(t)}(E_\infty^K)$$

by the Markov property,

$$= \psi(X(t)) \quad \text{a.e. } Q_x.$$

Next we show that $\psi(X(t))$ is a martingale relative to Q_x. For, if $s < t$,

$$E^{Q_x}(\psi(X(t))|\mathcal{F}_s)$$

$$= E^{Q_x}(E^{Q_x}(F|\mathcal{F}_t)|\mathcal{F}_s)$$

$$= E^{Q_x}(F|\mathcal{F}_s)$$

$$= \psi(X(s)).$$

Equating the expectations at time $t = 0$ and time t one gets

$$\psi(x) = \int_\Omega \psi(X(t))dQ_x$$

$$= \int \psi(y)\, q(t,x,y)dy,$$

where $q(t,x,A) = Q_x(X_t \in A)$, $\forall$ A Borel in $\mathbb{R}^d$.

We assume for the pr sent that $\psi(x)$ is continuous (This will be shown in Lemma 4 in the next section). By definition $0 \leq \psi \leq 1$.

Step 2. $\psi(x) = 1$, $\forall$ x or $\psi(x) = 0$, $\forall$ x.

Suppose that $\psi(x_0) = 0$ for some x_0. Then $0 = \psi(x_0) = \int \psi(y)\, q(t,x_0,y)\, dy$. As $q > 0$ a.e. and $\psi \geq 0$ we conclude that $\psi(y) = 0$ a.e. (with respect to Lebesgue measure). Since ψ is continuous ψ must vanish identically.

If $\psi(x_0) = 1$ for some x_0, we apply the above argument to $1 - \psi$ to conclude that $\psi = 1$, $\forall$ x. We now show that the third possibility $0 < \psi(x) < 1$ can never occur.

Since K is compact and ψ is continuous,

$$0 < a = \inf_{y \in K} \psi(y) \leq \sup_{y \in K} \psi(y) = b < 1.$$

From an Exercise in the section on martingales it follows that

$$\psi(X(t)) \to \chi_{E_\infty^K} \quad \text{a.e. } Q_x \quad \text{as } t \to +\infty .$$

Therefore $\lim_{t \to \infty} \psi(X(t))\,(1 - \psi(X(t))) = 0$ a.e. Q_x. Now

$$\psi(x_0) = Q_{x_0}(E_\infty^K) = Q_{x_0}\{w: w(t) \in K \text{ for arbitrary large time}\}$$

$$\leq Q_{x_0}\{w: a \leq \psi(X(t,w)) \leq b \text{ for arbitrarily large times}\}$$

$$\leq Q_{x_0}\{w: (1 - b)a \leq \psi(X(t))[1 - \psi(X(t)] \leq b(1-a) \text{ for arbitrarily large times}\}$$

$$= 0.$$

Thus $\psi(x) = 0$ identically, which is a contradiction. Thus for the given compact set K,

$$\text{either} \quad Q_x(E_\infty^K) = 0, \ \forall\, x,$$

$$\text{or} \quad Q_x(E_\infty^K) = 1, \ \forall\, x.$$

Step 3. If $Q_x(E_\infty^{K_0}) = 1$ for some compact set $K_0(\mathring{K}_0 \neq \emptyset)$ and $\forall\, x$, then $Q_x(E_\infty^K) = 1$, $\forall$ compact set K with non-empty interior.

We first given an intuitive argument. Suppose $Q_x(E_\infty^{K_0}) = 1$, i.e. almost all trajectories visit K_0 for arbitrarily large times. Each time a trajectory hits K_0, it has some chance of h itting K. Since the trajectory visits K_0 for arbitrarily large times it will visit K for arbitrarily large times. We now give a precise argument. Let

$$\tau_0 = \inf\{t: X(t) \in K_0\}$$

$$\tau_1 = \inf\{t \geq t_0 + 1 \quad X(t) \in K_0\}$$

$$\cdots \quad \cdots \quad \cdots \quad \cdots$$

$$\tau_n = \inf\{t \geq t_{n-1} + 1: X(t) \in K_0\}$$

$$\cdots \quad \cdots \quad \cdots \quad \cdots$$

Clearly $\tau_0 < \tau_1 < \ldots <$ and $\tau_n \geq n$.

$$Q_x(E_n^K) \geq Q_x \{X(t) \in K \text{ for } t \geq \tau_n\}$$

$$\geq Q_x \{X(t) \in K \text{ for } t \in \bigcup_{j=n}^{\infty} [\tau_j, \tau_j + 1]\}$$

$$= 1 - Q_x \{\bigcap_{j \geq n} X(t) \notin K \text{ for } t \in [\tau_j, \tau_j + 1]\}$$

$$\geq 1 - Q_x \{\bigcap_{j \geq n} X(\tau_j + 1) \in K\}$$

We claim that

$$Q_x(\bigcap_{j \geq n} X(\tau_j + 1) \notin K) = 0,$$

so that $Q_x(E_n^K) = 1$ for every n and hence $Q_x(E_\infty^K) = 1$, completing the proof of the theorem.

Now

$$q(1,x,K) \geq q(1,x,\overset{\circ}{K}) > 0, \forall x, \overset{\circ}{K} \text{ interior of } K.$$

It is clear that if $x_n \to x$, then

$$\underline{\lim}\ q(1,x_n,\overset{\circ}{K}) \geq q(1,x,\overset{\circ}{K}).$$

Let $d = \inf_{x \in K_0} q(1,x,\overset{\circ}{K})$. Then there exis ts a sequence x_n in K_0 such that $d = \operatorname{Lt}_{n \to \infty} q(1,x_n,\overset{\circ}{K})$. K_0 being compact, there exists a subsequence y_n of x_n with $y_n \to x$ in K_0, so that

$$d = \lim_{n \to \infty} q(1,x,\overset{\circ}{K}) = \underline{\lim}\ q(1,y_n,\overset{\circ}{K}) \geq q(1,x,\overset{\circ}{K}) > 0.$$

Thus

$$\inf_{x \in K_0} q(1,x,K) \geq d > 0.$$

Now

$$Q_x\Big(\ \ _{j=n}^{N}\ X(\tau_j+1) \notin K \,\Big|\, \mathcal{F}_{\tau_N}\Big)$$

$$= \prod_{j=n}^{N-1} \chi(X(\tau_j+1) \notin K)\ Q_x(X(\tau_N+1) \in K \,|\, \mathcal{F}_{\tau_N}) \quad \text{because}$$

$$\tau_j + 1 \leq \tau_N \quad \text{for} \quad j < N,$$

$$= \prod_{j=n}^{N-1} \chi(X(\tau_j+1) \notin K)\ Q_{X(\tau_N)}(X(1) \notin K) \quad \text{by the strong}$$

Markov property,

$$= \prod_{j=1}^{N-1} q(1, X(\tau_N), K^c)\ \chi(X(\tau_j+1) \notin K)\ .$$

Therefore

$$Q_x\Big(\bigcap_{j=n}^{N} X(\tau_j+1) \notin K\Big)$$

$$= E^{Q_x}\Big(Q_x\Big(\bigcap_{j=n}^{N} X(\tau_j+1) \notin K \,\Big|\ \ _{\tau_N}\Big)\Big)$$

$$= E^{Q_x}\Big(\prod_{j=n}^{N-1} \big(\chi_{[X(\tau_j+1) \notin K]}\big)\ q(1, X(\tau_N), K^c)\Big)$$

Since K_0 is compact and $X(\tau_N) \in K_0$,

$$q(1, X(\tau_N), K^c) = 1 - q(1, X(\tau_N), K) \leq 1 - d$$

Hence

$$Q_x\Big(\bigcap_{j=n}^{N} X(\tau_j+1) \notin K\Big) \leq (1-d)\, Q_x\Big(\bigcap_{j=n}^{N-1} X(\tau_j+1) \notin K\Big).$$

Iterating, we get

$$Q_x\Big(\bigcap_{j=n}^{N} X(\tau_j+1) \notin K\Big) \leq (1-d)^{N-n+1}, \ \forall\, N.$$

Let $N \to \infty$ to get

$$Q_x\Big(\ {}_{j=n}^{\infty} \ X(\tau_j + 1) \notin K\Big) = 0,$$

since $0 \leq 1\text{-}d < 1$. Thus the claim is proved and so is the theorem.

<u>COROLLARY</u>. *Let* K *be compact,* $\overset{\circ}{K} \neq \emptyset$ *Then* $Q_x(E_\infty^K) = 1$ *if and only if* $Q_x(E_0^K) = 1, \ \forall \ x$.

<u>Proof</u>. Suppose $Q_x(E_\infty^K) = 1$; then $Q_x(E_0^K) = 1$ because $E_\infty^K \quad E_0^K$. Suppose $Q_x(E_0^K) = 1$, then

$$\begin{aligned} Q_x(E_n^K) &= E^{Q_x}\big(E^{Q_x}(\chi_{E_n^K} \mid \mathcal{F}_n)\big) \\ &= E^{Q_x}\big(Q_{X(n)}(E_0^K)\big) \\ &= E^{Q_x}(1) \\ &= 1, \ \forall \ n. \end{aligned}$$

Therefore $Q_x(E_\infty^K) = 1$.

<u>REMARK</u>. If $Q_x(E_\infty^K) = 0$ then it need not imply that

$$Q_x(E_0^K) = 0.$$

<u>Example</u>. Take $b = 0$ and $d = 3$. Let $K = S_1 = \{x \in \mathbb{R}^3$ such that $|x| \leq 1\}$. Define

$$\psi(n) = \begin{cases} 1, & \text{for } |x| \leq 1, \\ \dfrac{1}{|x|}, & \text{for } |x| \geq 1. \end{cases}$$

$P_x(E_0^K) \neq$ constant but $P_x(E_\infty^K) = 0$. In fact, $P_x(E_0^K) = \psi(x)$ (Refer Dirichlet Problem).

31. INVARIANT PROBABILITY DISTRIBUTIONS

DEFINITION. Let $\{P_x\}_{x \in \mathbb{R}^d}$ be a family of Markov process on $\Omega = C([0,\infty); \mathbb{R}^d)$ indexed by the starting points x, with homogeneous transition probability $p(t,x,A) = P_x(X_t \in A)$ for every Borel set A in $\mathbb{R}^d$. A probability measure μ on the Borel field of $\mathbb{R}^d$ is called an *invariant distribution* if, $\forall$ A Borel in $\mathbb{R}^d$.

$$\int_{\mathbb{R}^d} p(t,x,A)\, d\mu(x) = \mu(A).$$

We shall denote $dp(t,x,y)$ by $p(t,x,dy)$ or $p(t,x,y)dy$ if it has a density.

PROPOSITION. *Let* $L_2 = \Delta/2 + b.\nabla$ *with no explosion. Let* Q_x *be the associated measure. If* $\{Q_x\}$ *has an invariant measure* μ *then the process is recurrent.*

Proof. It is enough to show that if K is a compact set with non-empty interior then

$$Q_x(E_\infty^K) = 1$$

for some x. Also $Q_x(E_t^K) \geq Q_x(X_t \in K) = q(t,x,K)$. Therefore

$$\mu(K) = \int q(t,x,K)\, d\mu(x) \leq \int Q_x(E_t^K)\, d\mu(x).$$

Now, $0 \leq Q_x(E_t^K) \leq 1$ and $Q_x(E_t^K)$ decreases to $Q_x(E_\infty^K)$ as $t \to \infty$. Therefore by the dominated convergence theorem

$$\mu(K) \leq \int Q_x(E_\infty^K)\, d\mu(x).$$

If the process were transient, then $Q_x(E_\infty^{S_n}) = 0, \forall n$, where $S_n = \{x \in \mathbb{R}^d : |x| \leq n\}$, i.e. $\mu(S_n) = 0, \forall n$. Therefore $\mu(\mathbb{R}^d) = 0$, which is false. Thus the process is recurrent.

The converse of this proposition is ***not*** true as is seen by the following example.

Let $L = \frac{1}{2}\frac{\partial^2}{\partial x^2}$ so that we are in a one-dimensional situation (Brownian motion). Then

$$p(t,x,K) = \int_K \frac{1}{\sqrt{(2\pi t)}} e^{\frac{-(x-y)^2}{2t}}\, dy \leq \frac{1}{\sqrt{(2\pi t)}} \lambda(K),$$

where λ denotes the Lebesgue measure on $\mathbb{R}$. If there exists an invariant distribution μ, then

$$\mu(K) = \int p(t,x,K)d\mu(x) \leq \frac{1}{\sqrt{(2\pi t)}} \lambda(K) \int d\mu(x) = \frac{\lambda(K)}{\sqrt{(2\pi t)}}$$

Letting $t \to \infty$, we get $\mu(K) = 0 \; \forall$ compact K, giving $\mu = 0$, which is false.

<u>THEOREM</u>. *Let* $L = \Delta/2 + b.\nabla$ *with no explosion. Assume* b *to be* C^∞. *Define the formal adjoint* L^* *of* L *by* $L^* = \Delta/2 - \nabla.b$ (i.e. $L^*u = \frac{1}{2}\Delta u - \nabla.(bu)$). *Suppose there exists a smooth function* ϕ (C^2 *- would do*) *such that* $L^*\phi = 0, \phi \geq 0, \int \phi\, dx = 1$. *If one defines* μ *by the rule* $\mu(A) = \int_A \phi(y)dy$, *then* μ *is an invariant distribution relative to the family* $\{Q_x\}$.

Proof. We assume the following result from the theory of partial differential equations.

If $f \in C_0^\infty(G)$ where G is a bounded open set with a smooth boundary ∂G and $f \geq 0$, then there exists a smooth function $U_G : [0,\infty) \times \overline{G} \to [0,\infty)$ such that

$$\frac{\partial U_G}{\partial t} = LU_G \quad \text{on} \quad (0,\infty) \times G,$$

$$U_G(0,x) = f(x) \quad \text{on} \quad \{0\} \times \overline{G},$$

$$U_G(t,x) = 0 \ , \ \forall \ x \in G.$$

Let $t > 0$. As U_G, ϕ are smooth and G is b ounded, we have

$$\frac{\partial}{\partial t} \int_G U_G(t,x) \ \phi(x)dx = \int_G \frac{\partial}{\partial t} U_G \phi \ dx = \int_G \phi LU_G dx$$

Using Green's formula this can be written as

$$\frac{\partial}{\partial t} \int_G U_G(t,x)\phi(x)dx = \int_G U_G L^*\phi - \frac{1}{2} \int_{\partial G} [\phi \frac{\partial U_G}{\partial n} - U_G \frac{\partial}{\partial n}] \ dS +$$

$$+ \int_{\partial G} < b.n > U_G(t,x)\phi(x)dS$$

Here n is assumed to be the inward normal to ∂G. So,

$$\frac{\partial}{\partial t} \int_G U_G(t,x)\phi(x)dx = - \frac{1}{2} \int_{\partial G} (x) \frac{\partial U_G}{\partial n} (t,x) \ dS$$

(Use the equation satisfied by ϕ and the conditions on U_G). Now $U_G(t,x) \geq 0$, $\forall \ x \in G$, $U_G(t,x) = 0$, $\forall \ x$ in ∂G, so that

$$\frac{\partial U_G}{\partial n} (t,x) \geq 0.$$

This means that

$$\frac{\partial}{\partial t}\int_G U_G(t,x)\ \phi(x)dx \le 0,\ \forall\ t > 0,$$

i.e. $\int_G U_G(t,x)\ \phi(x)dx$ is a monotonically decreasing function of t.
Therefore

$$\int_G U_G(t,x)\ \phi(x)dx \le \int_G U_G(0,x)\ \phi(x)dx$$

$$= \int_G f(x)\ \phi(x)dx$$

$$= \int_{\mathbb{R}^d} f(x)\ \phi(x)dx.$$

Next we prove that if $U: [0,\infty) \times \mathbb{R}^d \to [0,\infty)$ is such that $\frac{\partial U}{\partial t} = LU,\ \forall\ t > 0$ and $U(0,x) = f(x)$, then

$$\int_{\mathbb{R}^d} U(t,x)\ \phi(x)dx \le \int_{\mathbb{R}^d} f(x)\ \phi(x)dx.$$

The solution $U_G(t,x)$ can be obtained by using Itô calculus and is given by

$$U_G(t,x) = \int f(X(t))\ \chi_{\{\tau_G > t\}}\ dQ_x.$$

We already know that

$$U(t,x) = \int f(X(t))\ dQ_x.$$

Therefore

$$\int U(t,x)\ \phi(x)dx = \iint f(X(t))\ \phi(x)DQ_x\ dx.$$

Now

$$\int \int f(X(t))\, \chi_{\{\tau_G > t\}}\, dQ_x\, \phi(x)\, dx$$

$$\int U_G(t,x)\, \phi(x)dx \leq \int_{\mathbb{R}^d} f(x)\, \phi(x)dx.$$

Letting G increase to $\mathbb{R}^d$ and using Fatou's lemma, we get

$$\iint f(X(t))\, \phi(x)dQ_x dx \leq \int f(x)\, \phi(x)dx$$

This proves the assertion made above.

Let

$$\mu(A) = \int_A \phi(X)dx,$$

$$\nu(A) = \int Q_x(X_t \varepsilon A)d\mu(x) = \int q(t,x,A)d\mu(x).$$

Let $f \in C_0^\infty(G)$, $f \geq 0$, where G is a bounded open set with smooth boundary. Now

$$\begin{aligned} \int f(y)\, d\nu(y) &= \iint f(y)\, q(t,x,y)d\mu(x)dy \\ &= \iint f(X(t))dQ_x d\mu(x) \\ &= \int U(t,x)\, d\mu(x) \\ &= \int U(t,x)\, \phi(x)dx \\ &\leq \int f(x)\, \phi(x)dx = \int f(x)d\mu(x). \end{aligned}$$

Thus, $\forall\, f \geq 0$ such that $f \in C_0^\infty$,

$$\int f(x)\, d\nu(x) \leq \int f(x)d\mu(x).$$

This implies that $\nu(A) \leq \mu(A)$ for every Borel set A. (Use mollifier s and the dominated convergence theorem to prove the above inequality for χ_A when A is bounded). Therefore $\nu(A^c) \leq \mu(A^c)$, or $1 - \mu(A) \leq 1 - \mu(A)$, since μ,ν are both probability measures. This gives $\mu(A) = \nu(A)$, i.e.

$$\mu(A) = \int q(t,x,A)d\mu(x), \ \forall\ t,$$

i.e. μ is an invariant distribution.

We now see whether the converse result is true or not. Suppose there exists a probability measure μ on $\mathbb{R}^d$ such that

$$\int Q_x(X_t \in A)d\mu(x) = \mu(A), \ \forall\ A \text{ Borel in } \mathbb{R}^d \text{ and } \forall\ t.$$

The question we have in mind is whether $\mu(A) = \int_A \phi dx$ for some smooth ϕ satisfying $L^*\phi = 0$, $\phi \geq 0, \int \phi(x)dx = 1$. To answer this we proceed as follows.

By definition $\mu(A) = \int q(t,x,A)d\mu(x)$. Therefore

$$\iint f(X(t))\ dQ_x d\mu(x)$$

$$= \iint f(y)q(t,x,y)dyd\mu(x)$$

(1) $$= \int f(y)\ d\mu(y)\ \forall\ f \in C_0^\infty(\mathbb{R}^d)\ \|f\|_\infty \leq 1.$$

Since X is an Itô process relative to Q_x with parameters b and I,

$$f(X(t)) - \int_0^t (Lf)(X(s))ds$$

is a martingale. Equating the expectations at time $t = 0$ and time t we obtain

$$E^{Q_x}(f(X(t)) = f(x) + E^{Q_x}\left(\int_0^t (Lf)(X(s))ds\right)$$

Integrating this expression with respect to μ gives

$$\iint f(X(t))dQ_x d\mu(x) = \int f(x)d\mu(x) \int\int_{\mathbb{R}^d}\int_0^t (Lf)(X(s))ds dQ_x d\mu.$$

Using (1), we get

$$0 = \int_{\mathbb{R}^d}\int_{\Omega}\int_0^t (Lf)(X(s))dQ_x ds\, d\mu(x)$$

Applying equation (1) to the function Lf we then get

$$0 = \int_{\mathbb{R}^d}\int_0^t (Lf)(y)d\mu(y)ds$$

$$= t\int_{\mathbb{R}^d} (Lf)(y)d\mu(y),\ \forall\, t > 0.$$

Thus

$$0 = \int_{\mathbb{R}^d} (Lf)\,(y)\,d\mu(y),\ \forall\, f \in C_0^\infty(\mathbb{R}^d).$$

In the language of distributions this just means that $L^*\mu = 0$.

From the theory of partial differential equations it then follows that there exists a smooth function ϕ such that $\forall\, A$ Borel in $\mathbb{R}^d$,

$$\mu(A) = \int_A \Phi(y)dy$$

with $L^*\phi = 0$. As $\mu \geq 0$, $\phi \geq 0$ and since

$$\mu(\mathbb{R}^d) = 1, \int_{\mathbb{R}^d} \phi(x)\, dx = 1.$$

We have thus proved the following (converse of the previous) theorem.

THEOREM. *Let* μ *be an invariant distribution with respect to the family* $\{Q_x\}$ *with* $b: \mathbb{R}^d \to \mathbb{R}^d$ *being* C^∞. *Then there exists a* $\phi \in L^1(\mathbb{R}^d)$, $\phi \geq 0, \phi$ *smooth such that*

$$L^*\phi = 0, \int \phi(y)\, dy = 1$$

and such that

$$\mu(A) = \int_A \phi(y)\, dy, \; \forall A \textit{ Borel in } \mathbb{R}^d.$$

THEOREM (UNIQUENESS). *Let* ϕ_1, ϕ_2 *be smooth on* $\mathbb{R}^d$ *such that*

$$\phi_1, \phi_2 \geq 0, \; 1 = \int_{\mathbb{R}^d} \phi_1 dy = \int_{\mathbb{R}^d} \phi_2 dy, \; L^*\phi_1 = 0 = L^*\phi_2.$$

Then $\phi_1 = \phi_2$.

Proof. Let $f(x) = \phi_1(x) - \phi_2(x)$,

$$\mu_i(A) = \int_A \phi_i(x) dx, \; i = 1,2.$$

Then μ_1, and μ_2 are invariant distributions. Therefore

$$\int q(t,x,y)\, \phi_i(x) dx = \int q(t,x,y)\, d\mu_i(x)$$

$$= \phi_i(y), \text{ (a.e.)}, \; i = 1,2.$$

Taking the difference we obtain

$$\int q(t,x,y)\, f(x)\, dx = f(y), \text{ a.e.}$$

Now

$$\int |f(y)\ dy = \int|\int q(t,x,y)\ f(x)dx|\ dy$$

$$\leq \iint q(t,x,y)\ |f(x)|\ dxdy$$

$$= \int |f(x)|\ dx \int q(t,x,y)dy$$

$$= \int |f(x)|\ dx.$$

Thus

$$(*) \qquad \iint|f(x)|\ q(t,x,y)dxdy = \int|\int q(t,x,y)f(x)dx|\ dy \quad \forall\ t.$$

We show that f does not change sign, i.e. $f \geq 0$ a.e. or $f \leq 0$ a.e. The result then follows from the fact that $\int f(x)dx = 0$. Now

$$|\int q(1,x,y)\ f(x)dx| \leq \int q(1,x,y)|f(x)|\ dx$$

and (*) above gives

$$\int|\int q(1,x,y)\ f(x)dx|dy = \iint q(1,x,y)|f(x)|dxdy.$$

Thus

$$|\int q(1,x,y)f(x)dx| = \int q(1,x,y)|f(x)|dx \quad \text{a.e. } y,$$

i.e.

$$|\int_{E^-} q(1,x,y)f(x)dx + \int_{E^-} q(1,x,y)\ f(x)dx|$$

$$= \int_{E^+} q(1,x,y)f(x)dx - \int_{E^-} q(1,x,y)f(x)dx \quad \text{a.e. } y,$$

where

$$E^+ = \{x\colon f(x) > 0\},\ E^- = \{x\colon f(x) < 0\},\ E^0 = \{x\colon f(x) = 0\}.$$

Squaring both sides of the above equality, we obtain

(**) $\left(\int_{E^+} q(1,x,y)\, f(x)dx\right) \left(\int_{E^-} q(1,x,y)f(x)dx\right) = 0$, a.e. y.

Let A be a set of positive Lebesgue measure; then $p(1,x,A) = P_x(X(1) \in A) > 0$. Since Q_x is equivalent to P_x on Ω we hhave $Q_x(X(1) \in A) = q(1,x,A) > 0$. Therefore $q(1,x,y) > 0$ a.e. y for each x. By Fubini's theorem $q(1,x,y) > 0$ a.e. x,y. Therefore for almost all y, $q(1,x,y) > 0$ for almost all x. Now pick a y such that (**) holds for which $q(1,x,y) > 0$ a.e. x.

We therefore conclude from (**) th at either

$$\int_{E^+} q(1,x,y)f(x)dx = 0, \text{ in which case } f \leq 0 \text{ a.e.,}$$

or

$$\int_{E^-} q(1,x,y)f(x)dx = 0, \text{ in which case } f \geq 0 \text{ a.e.}$$

Thus f does not change its sign, which completes the proof.

REMARK. The only property of the operator L we used was to conclude $q > 0$. We may therefore expect a similar result for more general operators.

THEOREM. *Let* $L^*\phi = 0$ *where* $\phi \geq 0$ *is smooth and* $\int \phi(x)dx = 1$. *Let* K *be any compact set. Then*

$$\sup_{x \in K} \int |q(t,x,y) - \phi(y)|\, dy \to 0 \text{ as } t \to +\infty.$$

LEMMA 1. *Let* b *be bounded and smooth. For every* $f: \mathbb{R}^d \to \mathbb{R}^d$ *that is bounded and measurable let* $u(t,x) = E^{Q_x}(f(X(t))$. *Then for every fixed* $t, u(t,x)$ *is a continuous function of* x *Further, for* $t \geq \varepsilon > 0$,

$$\left| u(t,x) - \int u(t-\varepsilon,y) \frac{1}{\sqrt{(2\pi\varepsilon)}^d} \exp - \frac{|x-y|^2}{2\varepsilon} \, dy \right|$$

$$\leq \| f \|_\infty \sqrt{(e^{ct}(e^{c\varepsilon}-1))},$$

where c *is a constant depending only on* $\| b \|_\infty$.

Proof. Let

$$(T_t f)(x) = E^{Q_x}(f(X(t)) = E^{P_x}(f(X(t))Z(\varepsilon,t)) +$$

(1)

$$+ E^{P_x}(f(X(t)) (Z(t)-Z(\varepsilon,t))),$$

where

$$Z(t) = \exp \left[\int_0^t < b; dx > - \frac{1}{2} \int_0^t |b|^2 ds \right],$$

$$Z(\varepsilon,t) = \exp \left[\int_\varepsilon^t < b,dx > - \frac{1}{2} \int_\varepsilon^t |b (X(s))|^2 ds \right].$$

$$E^{P_x}(f(X(t)) Z(\varepsilon,t)) = E^{P_x}(E^{P_X}(f(X(t)) Z(\varepsilon,t)|_\varepsilon))$$

$$= E^{P_x}(E^{P_{X_\varepsilon}})(f(X(t-\varepsilon))Z(t-\varepsilon)))$$

(by Markov property),

$$= E^{P_x}(u(t - \varepsilon, X(\varepsilon)).$$

(2) $$= \int u(t-\varepsilon,y) \frac{1}{(\sqrt{(2\pi\varepsilon)})^d} \exp \left[- \frac{|(x-y)|^2}{2\varepsilon}\right] dy.$$

Now

$$(E^{P_x}(|Z(t) - Z(\varepsilon,t)|)) =$$

$$= E^{P_x}(|Z(\varepsilon) Z(\varepsilon,t) - Z(\varepsilon,t)|))^2$$

$$= E^{P_x}(Z(\varepsilon,t) Z(\varepsilon) - 1|))^2$$

$$\leq (E^{P_x}((Z(\varepsilon)-1)^2))(E^{P_x}(Z^2(\varepsilon,t)))$$

(by Cauchy Schwarz inequality),

$$\leq E^{P_x}(Z^2(\varepsilon) - 2Z(\varepsilon) + 1)\, E^{P_x}(Z^2(\varepsilon,t))$$

$$\leq E^{P_x}(Z^2(\varepsilon)-1)E^{P_x}(Z^2(\varepsilon,t)), \text{(since } E^{P_x}(Z(\varepsilon)) = 1),$$

$$\leq E^{P_x}(Z^2(\varepsilon)-1)E^{P_x}(\exp(2\int_\varepsilon^t \langle b,dX\rangle - \frac{2}{2}\int_\varepsilon^t |b|^2 ds + \int_\varepsilon^t |b|^2 ds))$$

$$\leq E^{P_x}(Z^2(\varepsilon) - 1)e^{ct},$$

using Cauchy Schwarz inequality and the fact that

$$E^{P_x}(\exp(2\int_\varepsilon^t \langle b,dX\rangle - \frac{2^2}{2}\int_\varepsilon^t |b|^2 ds)) = 1.$$

Thus

$$E^{P_x}(|Z(t) - Z(\varepsilon, t)|)^2 \leq (e^{c\varepsilon} - 1)e^{ct}$$

where c depends only on $\|b\|_\infty$. Hence

$$|E^{P_x}(f(X(t))(Z(t) - Z(\varepsilon,t))| \leq \|f\|_\infty . E^{P_x}(|Z(t)- Z(\varepsilon,t)|)$$

(3) $$\leq \|f\|_\infty \sqrt{((e^{c\varepsilon} - 1)e^{ct})}.$$

Substituting (2) and (3) in (1) we get

$$|u(t,x) - \int u(t-\varepsilon y).\frac{1}{(\sqrt{(2\pi\varepsilon)})^d} \exp[\frac{-|x-y|^2}{2\varepsilon}]dy$$

$$\leq \|f\|_\infty \sqrt{((e^{c\varepsilon} - 1)e^{ct})}$$

Note that the right hand side is independent of x and as $\varepsilon \to 0$ the right hand side converges to 0. Thus to show that $u(t,x)$ is a continuous function of x (t fixed), it is enough to show that

$$\int u(t-\varepsilon,y) \frac{1}{(\sqrt{(2\pi\varepsilon)})^d} \exp[\frac{-|x-y|^2}{2\varepsilon}]dy$$

is a continuous function of x; but this is clear since u is bounded. Thus for any fixed t $u(t,x)$ is continuous.

LEMMA 2. *For any compact set* $K \subset R^d$, *for* r *large enough so that*

$$K \subset \{x: |x| < r\}, \; x \to Q_x(\tau_r \leq t)$$

is continuous on K *for each* $t \geq 0$, *where*

$$\tau_r(w) = \inf \{s: |w(s)| \geq r\}.$$

Proof. $Q_x(\tau_r \leq t)$ depends only on the coefficient $b(x)$ on $|x| \leq r$. So modifying, if necessary, outside $|x| \leq r$, we can very well assume that $|b(x)| \leq M$ for all x. Let

$$\tau_r^\varepsilon = \inf \{s: s \geq \varepsilon, |w(s)| \geq r\}.$$

$$Q_x(\tau_r^\varepsilon \leq t) = E^{Q_x}(u(X(\varepsilon))),$$

where

$$u(x) = Q_x(\tau_r \leq t - \varepsilon)$$

As b and u are bounded, for every fixed $\varepsilon > 0$, by Lemma 1, $Q_x(\tau_r \leq t)$ is a continuous function of x. As

$$|Q_x(\tau_r^\varepsilon \leq t) - Q_x(\tau_r \leq t)| \leq Q_x(\tau_r \leq \varepsilon),$$

to prove the lemma we have only to show that

$$\underset{\varepsilon \to 0}{\text{limit}} \; \sup_{x \in K} \; Q_x(\tau_r \leq \varepsilon) = 0$$

Now

$$Q_x(\tau_r \leq \varepsilon) = \int_{\{\tau_r \leq \varepsilon\}} Z(\;)\, dP_x$$

$$\leq (\int (Z(\varepsilon))^2 \, dP_x)^{1/2} \cdot \sqrt{P_x(\tau_r \leq \varepsilon)},$$

by Cauchy-Schwarz inequality. The first factor is bounded because b is bounded. The second factor tends to zero uniformly on K because

$$\sup_{x \in K} P_x(\tau_r \leq \varepsilon) \leq P\left(\sup_{0 \leq s \leq \varepsilon} |w(s)| > \delta\right)$$

where

$$\delta = \inf_{\substack{y \in K \\ |x| = r}} |(x-y)|.$$

<u>LEMMA 3</u>. *Let* K *be compact in* R^d. *Then for fixed* t, $Q_x(\tau_r \leq t)$ *monotically decreases to zero as* $r \to \infty$ *and the convergence is uniform on* K.

<u>Proof</u>. Let $f_r(x) = Q_x(\tau_r \leq t)$. As $\{\tau_r \leq t\}$ decreases to the null set, $f_r(x)$ decreases to zero. As K is compact, there exists an r_0 such that for $r \geq r_0$, $f_r(x)$ is continuous on K, by Lemma 2. Lemma 3 is a consequence of Dini's theorem.

<u>LEMMA 4</u>. *Let* $b: R^r \to R^d$ *be smooth (not necessarily bounded). Then* $E^{Q_x}(f(X(t)))$ *is continuous in* x *for every fixed* t, f *being any bounded measurable function.*

<u>**Proof.**</u> Let b_r be any bounded smooth function on R^d such that $b_r \equiv b$ on $|x| \leq r$ and Q_x^r the measure corresponding to b_r. Then by Lemma 1, $E^{Q_x^r}(f(X(t)))$ is continuous in x for all r. Further,

$$|E^{Q_x^r}(f(X(t))) - E^{Q_x}(f(X(t)))| \leq 2 \|f\|_\infty \cdot Q_x(\tau_r \leq t).$$

The result follows by Lemma 3.

LEMMA 5. *With the hypothesis as the same as in Lemma 1,* (S_1) *is an equicontinuous family, where*

$$S_1 = \{f: \mathbb{R}^d \to \mathbb{R}, f \text{ bounded measurable}, \|f\|_\infty \le 1\}$$

Proof. For any f in S_1, let $U(x) = U(t,x) \equiv E^{Q_x}(f(X(t))$ and

$$U_\varepsilon(x) = U_\varepsilon(t,x) = \int U(t-\varepsilon,y)\frac{1}{(\sqrt{(2\pi\varepsilon)})^d} \exp [\frac{-|x-y|^2}{2\varepsilon}]dy.$$

By Lemma 1,

$$|U(x) - U_\varepsilon(x)| \le (((e^{c\varepsilon}-1)e^{ct}))^{1/2}$$

$$|U(x)-U(y)| \le |U(x)-U_\varepsilon(x)| + |U_\varepsilon(y)-U(y)| + |U_\varepsilon(x) -U_\varepsilon(y)|$$

$$\le 2\sqrt{((e^{c\varepsilon}-1)e^{ct})} + |U_\varepsilon(x)-U_\varepsilon(y)|.$$

The family $\{U_\varepsilon: f \in S_1\}$ is equicontinuous because every U occuring in the expression for U_ε is bounded by 1, and the exponential factor is uniformly continuous. Thus the right hand side is very small if ε is small and $|x-y|$ is small. This proves the lemma.

LEMMA 6. *Let* b *be smooth and assume that there is no explosion* (b *is not necessarily bounded). Then* (S_1) *is an equi-continuous family* $\forall\, t > 0$.

Proof. Let $r > 0$ be given. Define $b_r \in C^\infty$ such that $b_r = 0$ on $|x| > r+1$, $b_r = b$ on $|x| \le r$, $b_r: \mathbb{R}^d \to \mathbb{R}$. By Lemma 2, we have that

$$\{E^{Q_x^r}(f(X(t))): f \in S_1\}$$

is equicontinuous, where Q_x^r is the probability measure corresponding to the function b_r.

(1) $$E^{Q_x}(f(X(t))\chi_{\{\tau_r > t\}}) \; E^{Q_x^r}(f(X(t)) \; \chi_{\{\tau_r > t\}}).$$

Therefore

$$|E^{Q_x}(f(X(t))) - E^{Q_x^r}(f(X(t))|$$

$$= |E^{Q_x}(f(X(t))\chi_{\{\tau_r > t\}}) + E^{Q_x}(f(X(t))\chi_{\{\tau_r \leq t\}})$$

$$- E^{Q_x^r}(f(X(t))\chi_{\{\tau_r > t\}}) - E^{Q_x^r}(f(X(t)))\chi_{\{\tau_r \leq t\}})|$$

$$= |E^{Q_x}(f(X(t)\chi_{\{\tau_r \leq t\}}) - E^{Q_x^r}(f(X(t))\chi_{\{\tau_r \leq t\}})|$$

$$\leq \|f\|_\infty (E^{Q_x}(\chi_{\{\tau_r \leq t\}}) + E^{Q_x^r}(\chi_{\{\tau_r \leq t\}})$$

$$\leq 1[E^{Q_x}(\chi_{(\tau_r \leq t)}) + E^{Q_x}(\chi_{(\tau_r \leq t)})] \text{ (use (1) with } f = 1)$$

$$= 2E^{Q_x}(\chi_{(\tau_r \leq t)}).$$

Thus

$$\sup_{x \in K} \sup_{\|f\|_\infty \leq 1} |E^{Q_x}(f(X(t))-E^{Q_x^r}(f(X(t))| \leq 2 \sup_{x \in K} (\chi_{\{\tau_r \leq t\}}).$$

By Lemma 3,

$$\sup_{x \in K} E^{O_x}(\tau_r \leq t) \to 0$$

for every compact set K as $n \to \infty$, for every fixed t.

The equicontinuity of the family (S_1) now follows easily.

For fixed x_0, put $u_r(x) = E^{Q_x^r}(f(X(t))$ and $u(x) = E^{Q_x}(f(X(t))$ and let $K = S[x_0,1] = \{x: |x - x_0| \leq 1\}$. Then

$$|u(x)-u(x_0)| \leq |u(x)-u_r(x)| + |u(x_0)-u_r(x_0)| + |u_r(x)-u_r(x_0)|$$

$$\leq 2 \sup_{y \in K} E^{Q_y}(\chi_{(\tau_r \leq t)}) + |u_r(x)-u_r(x_0)|$$

By the previous lemma $\{u_r\}$ is an equicontinuous family and since $\sup_{y \in K} E^{Q_y}(\chi_{(\tau_r \leq 1)}) \to 0$, $\{u: \|f\|_\infty \leq 1\}$ is equicontinuous at x_0. This proves the Lemma.

<u>LEMMA 7</u>. $T_t \circ T_s = T_{t+s}, \ \forall\, s,t \geq 0.$

<u>REMARK</u>. This property is called the *semigroup property*.

<u>Proof</u>. $T_t(T_s f)(x)$

$$= \iint f(z)\, q(s,y,z)\, q(t,x,y)\, dydz.$$

Thus we have only to show that

$$\int q(t,x,y)\, q(s,y,A)dy = q(t+s,x,A).$$

$$q(t+s,x,A) = E^{Q_x}(X(t+s) \in A)$$

$$= E^{Q_x}(X(t+s) \in A| \ _t))$$

$$= E^{Q_x}(E^{Q_x} X(t)\ (X(s) \in A))),$$

by Markov property

$$= E^{Q_x}(q(s,X(t),A))$$

$$= \int q(t,x,y)\, q(s,y,A)dy,$$

which proves the result.

As a trivial consequence we have the following.

LEMMA 8. *Let* $\varepsilon > 0$ *and let* S_1 *be the unit ball in* $B(\mathbb{R}^d)$. *Then* $\bigcup_{t \geq \varepsilon} T_t(S_1)$ *is equicontinuous.*

Proof. $\bigcup_{t \geq \varepsilon > 0} T_t(S_1) = T\left(\bigcup_{t \geq 0} T_t(S_1)\right)$ (by Lemma 7) $T_\varepsilon(S_1)$.

The result follows by Lemma 6.

LEMMA 9. *Let* $u(ttx) = E^{Q_x}(f(X(t))$ *with* $\|f\|_\infty \leq 1$. *Let* $\varepsilon > 0$ *be given and* K *any compact set. Then there exists a* $T_0 = T_0(\varepsilon, K)$ *such that* $\forall\, T \geq T_0$ *and* $\forall\, x_1, x_2 \in K$,

$$|u(T,x_1) - u(T,x_2)| \leq \varepsilon.$$

Proof. Define $q^*(t,x_1,x_2,y_1,y_2) = q(t,x_1,y_1)\, q(t,x_2,y_2)$ and let $Q_{(x_1,x_2)}$ be the measure corresponding to the operator

$$L = \frac{1}{2}(\Delta_{x_1} + \Delta_{x_2}) + b(x_1).\nabla_{x_1} + b(x_2).\,\nabla_{x_2}$$

i.e., for any $u\colon \mathbb{R}^d \times \mathbb{R}^d \to \mathbb{R}$,

$$Lu = \frac{1}{2}\sum_{i=1}^{2d} \frac{\partial^2 u}{\partial x_i^2} + \sum_{i=1}^{d} b_i(x_1,\ldots,x_d)\,\frac{\partial u}{\partial x_i^2} +$$

$$+ \sum_{i=1}^{d} b_i(x_{d+1},\ldots,x_{2d})\,\frac{\partial u}{\partial x_{i+d}}\,.$$

Then $Q_{(x_1,x_2)}$ will be a measure on $C([0,\infty);\mathbb{R}^d \times \mathbb{R}^d)$. We claim that $Q_{(x_1,x_2)} = Q_{x_1} \times Q_{x_2}$. Note that

$$C([0,\infty); \mathbb{R}^d \times \mathbb{R}^d) = C([0,\infty); \mathbb{R}^d) \times C([0,\infty); \mathbb{R}^d)$$

and since $C([0,\infty); \mathbb{R}^d)$ is a second countable metric space, the Borel field of $C([0,\infty) \mathbb{R}^d \times \mathbb{R}^d)$ is the σ-algebra generated by $\mathcal{B} : (C([0,\infty); \mathbb{R}^d)) \times \mathcal{B}(C[0,\infty); \mathbb{R}^d)$.

By going to the finite-dimensional distributions one can check that $P_{(x_1,x_2)} = P_{x_1} \times P_{x_2}$.

$$\frac{dQ_{(x_1,x_2)}}{dP_{(x_1,x_2)}}\bigg|\mathcal{F}_t = \exp\left[\int_0^t < b^{(1)}, dX_1 > - \frac{1}{2}\int_0^t |b^{(1)}|^2 ds\right] \times$$

$$\times \exp\left[\int_0^t < b^{(2)}, dX_2 > - \frac{1}{2}\int_0^t |b^{(2)}|^2 ds\right],$$

where $b^{(1)}(x_1 \ldots x_d) = b(x_1 \ldots, x_d)$. $b^{(2)}(x_{d+1} \ldots x_{2d}) = b(x_{d+1}, \ldots, x_{2d})$, so that $Q_{(x_1,x_2)} = Q_{x_1} \times Q_{x_2}$.

It is clear that if ϕ defined an invariant measure for the process Q_x, i.e.

$$\int_A \phi(x)dx = \int \phi(y) Qy(X_t \in A)dy,$$

then $\phi(y_1)\,\phi(y_2)$ defines an invariant meas ure for the process $Q_{(x_1,x_2)}$. Thus the process $Q_{(x_1,x_2)}$ is recurrent.

Next we show that $u(T - t, X_1(t))$ is a martingale $(0 \leq t \leq T)$ for any fixed T on $C([0,T]; \mathbb{R}^d)$.

$$E^{Q_x}(u(T-t, X(t) \mid \mathcal{F}_s))$$

$$= \left[\int u(T-t,y)\, q(t-s,x,dy)\right]_{x = X(s)}$$

$$= [\iint f(z)\, q(T-t,y,dz)\, q(t-s,x,dy)]_{x = X(s)}$$

$$= [\int f(z)\, q(T-s,x,dz)]_{x = X(s)}$$

$$= u(T-s,X(s)),\ s < t.$$

It now follows that $u(T-t,X_1(t))$ is a martingale on $C([0,\infty);\mathbb{R}^d) \times C([0,\infty);\mathbb{R}^d)$. Hence $u(T-t,X_1(t)) - u(T-t,X_2(t))$ is a martingale relative to $Q_{(X_1,x_2)}$.

Let $V = S(0,\delta/2) \subset \mathbb{R}^d \times \mathbb{R}^d$ with $\delta < 1/4$. If $(x_1,x_2) \in V$, then

$$|x_1-x_2| \leq |(x_1,0) - (0,0)| + |(0,0) - (0,x_2)| < \delta.$$

<u>Claim 1</u>. $Q_{(x_1,x_2)}(\tau_V \leq T) \to 1$ as $T \to \infty$, where τ_V is the exit time from $R^d - V$.

<u>Proof.</u> If w is any trajectory starting at some point in V, then $\tau_V = 0 \leq T$, $\forall$ T. If w starts at some point outside V then, by the recurrence property, w has to visit a ball with centre 0 and radius $\delta/2$; hence it must get into V at some finite time. Thus $\{\tau_V \leq T\} \uparrow$ to the whole space as $T \uparrow \infty$. Next we show that the convergence is uniform on compact sets.

If $x_1,x_2 \in K$, $(x_1,x_2) \in K \times K$ (a compact set). Put $g_T(x_1,x_2) = Q_{(x_1,x_2)}(\tau_V \leq T)$. Then $g_T(x_1,x_2) \geq 0$ and $g_T(x_1,x_2)$ increases to 1 as T tends to ∞.

$$g_T(x_1,x_2) = Q_{(x_1,x_2)}(\tau_V \leq T)$$

$$Q_{(x_1,x_2)}\,(\tau_V^1 \leq T),$$

where

$$\tau_V^1 = \inf\{t \geq 1: (x_1,x_2) \in V\}.$$

Therefore

$$g_T(x_1,x_2) \geq E^Q(x_1,x_2)(E^Q(x_1,x_2)((\tau_V^1 \leq T)| \quad {}_1))$$

$$= E^Q(x_1,x_2)(Q_{(X_1(1),X_2(1))}\{\tau_V^1 \leq T)\})$$

$$= E^Q(x_1,x_2)(\psi_T(X_1(1),X_2(1))),$$

where ψ_T is a bounded non-negative function. Thus, if

$$h_T(x_1,x_2) = Q_{(x_1,x_2)}\,(\tau_V^1 \leq T) =$$

$$= E^Q(x_1,x_2)\,(\psi_T(X_1(1),X_2(1))),$$

then by Lemma 4, h_T is continuous for each T, $g_T \geq h_T$ and h_T increases to 1 as $T \to \infty$. Therefore, h_T converges uniformly (and so does g_T) on compact sets.

Thus given $\varepsilon > 0$ chose $T_0 = T_0(\varepsilon,K)$ such that if $T \geq T_0$,

$$\sup_{x_2 \in K}\ \sup_{x_1 \in K}\ Q_{(x_1,x_2)}(\tau_V \geq T-1) \leq \varepsilon.$$

By Doob's optional stopping theorem and the fact that $u(T-t,X_1(t)) - u(t-t,X_2(t))$ is a martingale, we get, on equating expectations,

$$|u(T,x_1) - u(T,x_2)|$$

$$= |E^{Q_{(x_1,x_2)}}[u(T-0,X_1(0) - u(T-0,X_2(0)]|$$

$$= |E^{Q_{(x_1,x_2)}}[u(T-(\tau_v \wedge(T-1)), X_1(T-(\tau_v \wedge (T-1)) -$$

$$- u(T-(\tau_v \wedge (T-1)), X_2(T-(\tau_v \wedge (T-1)]|$$

$$= |\int_{\{\tau_v \geq T-1\}} [u(1,X_1(1)) - u(1,X_2(1))]dQ_{(x_1,x_2)} +$$

$$+ \int_{\{\tau_v <(T-1)\}} [u(T-\tau_v,X_1(T-\tau_v))-u(T-\tau_v),X_2(T-\tau_v))dQ_{(x_1,x_2)}|.$$

Therefore

$$|u(T,x_1) - u(T,x_2)|$$

$$\leq \int_{\{\tau_v \geq(T-1)\}} |[u(1,X_1(1))-u(1,X_2(1))]|dQ_{(x_1,x_2)} +$$

$$+ |\int_{\{\tau_v < (T-1)\}} [u(T-\tau_v,X_1(T-\tau_v))-u(T-\tau_v,X_2(T-\tau_v))dQ_{(x_1,x_2)}|$$

$$\leq 2\varepsilon + |\int_{\{\tau_v < (T-1)\}} [u(T-\tau_v,X_1(T-\tau_v))-u(T-\tau_v,X_2(T-\tau_v))]dQ_{(x_1,x_2)}|,$$

since u is bounded by 1.

The second integration is to be carried out on the set $\{T-\ _v \geq 1\}$. Since $\bigcup_{t \geq 1} T_t(S_1)$ is equicontinuous we can choose a $\delta > 0$ such that whenever $x_1,x_2 \in K$ such that $|x_1 - x_2| < \delta$

$$|u(t,x_1) - u(t,x_2)| \le \varepsilon, \ \forall\ t \ge 1.$$

Thus $|u(T,x_1) - u(T,x_2)| \le 3\varepsilon$ whenever $x_1, x_2 \in K$ and $T \ge T_0$. This proves the Lemma.

COROLLARY TO LEMMA 9. $\sup\limits_{x_1,x_2 \in K} \int |q(t,x_1,y)|dy$ *converges to* 0 *as* $t \to \infty$.

Proof. Since the dual of L^1 is L^∞, we have

$$\int |q(t,x_1,y) - q(t,x_2,y)|dy$$

$$= \sup_{||f||_\infty \le 1} \left|\int [q(t,x_1,y) - q(t,x_2,y)]\, f(y)dy\right|$$

and the right side converges to 0 as $t \to \infty$, by Lemma 9.

We now come to the proof of the main theorem sstated before Lemma 1. Now

$$\int |q(t,x,y) - \phi(y)|dy$$

$$= \int |q(t,x,y) - \int \phi(x^1)q(t,x^1,y)dx^1|dy$$

(by invariance property)

$$= \int \left|\int q(t,x,y)\phi(x^1)dx^1 - \int \phi(x^1)q(t,x^1,y)dx^1\right|dy$$

$$\left(\text{since } \int \phi(x^1)dx^1 = 1\right)$$

$$\le \iint |q(t,x,y)-q(t,x^1,y)|\phi(x^1)dx^1dy \quad (\text{since } \phi \ge 0)$$

$$= \int \phi(x^1)dx^1 \int |q(t,x,y) - q(t,x^1,y)|dy$$

Since

$$\int \phi(x^1)dx^1 = \operatorname*{Lt}_{n\to\infty} \int_{|x^1| \le n} \phi(x^1)dx^1,$$

choose a compact set L K such that $\int_{\mathbb{R}^d - L} \phi(x^1)dx^1 < \varepsilon$. Then

$$\int \phi(x^1)dx^1 \int |q(t,x,y) - q(t,x^1,y)|dy$$

$$= \int_L \phi(x^1)dx^1 \int |q(t,x,y) - q(t,x^1,y)|dy +$$

$$+ \int_{\mathbb{R}^d - L} \phi(x^1)dx^1 \int |q(t,x,y) - q(t,x^1,y)|dy$$

$$\leq \int\int_L \phi(x^1)dx^1 \int |q(t,x,y) - q(t,x^1,y)|dy + 2\varepsilon.$$

Chose t_0 such that whenever $t \geq t_0$,

$$\int |q(t,x,y) - q(t,x^1,y)|dy \leq \frac{\varepsilon}{1 + \int_L \phi(x^1)dx^1}$$

$\forall\ x, x_1$ in L. (Corollary to Lemma 9). Then

$$\int |q(t,x,y) - \phi(y)|dy \leq 3\varepsilon,$$

if $t \geq t_0$ $\forall\ x \in K$ completing the proof of the theorem.

32. ERGODIC THEOREM

THEOREM. *Let* $f\colon \mathbb{R}^d \to R$ *be bounded and measurable with* $\|f\|_\infty \le 1$. *If* ϕ *is an invariant distribution for the family* $\{Q_x\}$, $x \in \mathbb{R}^d$ *then*

$$\lim_{\substack{t_1 \to \infty \\ 0 \le t_2 - t_1 \to \infty}} E^{Q_x}(f(X(t_1))f(X(t_2))) = [\int f(y)\,\phi(y)dy]^2$$

Proof. $E^{Q_x}[(f(X(t_1)) f(X(t_2))]$

$$= E^{Q_x}(E^{Q_x}[f(X(t_1))\, f(X(t_2))|\,\mathcal{F}_{t_1}])$$

$$= E^{Q_x}(f(X(t_1))\,(E^{Q_x}[f(X(t_2))|\,\mathcal{F}_{t_1}]))$$

$$= E^{Q_x}(f(X(t_1)) \int f(y)\, q(t_2 - t_1, X(t_1), y))dy), t_2 > t_1$$

(by Markov property),

(1) $$= \int f(z)\, q(t_1,x,z)dz \int f(y)\, q(t_2 - t_1, z, y)dy$$

doe any bounded and measurable f. By theorem of § 31,

$$\sup_{x \in K} \left|\int f(y)\, [q(t,x,y) - \phi(y)]\, dy\right| \to 0$$

as $t \to +\infty$. We can therefore write (1) in the form

$$E^{Q_x}[(f(X(t_1))\, f(X(t_2))] =$$

$$= (\int f(z)q(t_1,x,x)dz)\int f\phi + \int f(z)q(t_1,x,z)A(t_2-t_1,z)dz,$$

where $A(t_2 - t_1, z)$ converges to 0 (uniformly on compact sets as) $t_2 - t_1 \to +\infty$.

To prove the theorem we have therefore only to show that

$$\int f(z)\, q(t_1,x,z)\, A(t_2 - t_1,z)dz \to 0$$

as $t_1 \to +\infty$ and $t_2 - t_1 \to \infty$ (because $\int f(z)q(t_1,x,z)dz \to \int f\phi$).

Now

$$|\int f(z)\, q(t_1,x,z)\, A(t_2 - t_1,z)dz|$$

$$\leq \|f\|_\infty \int q(t_1,x,z)|A(t_2 - t_1,z)|dz$$

$$\leq \int q(t_1,x,z)|A(t_2 - t_1,z)|dz \tag{2}$$

Let K be any compact set, then

$$\int_K q(t_1,x,z)\, dz = \int \chi_K\, q(t_1,x,z)\, dz \to \int \chi_K\, \phi(z)dz$$

as $t_1 \to \infty$. Given $\varepsilon > 0$, let K be compact so that

$$|\int \chi_{K^c}\, \phi(z)\, dz| \leq \varepsilon;$$

then $|\int \chi_{K^c}\, q(t_1,x,z)dz| \leq 2\varepsilon$ if $t_1 >> 0$. Using (2) we therefore get

$$|\int f(z)\, q(t_1,x,z)\, A(t_2 - t_1,z)\, dz|$$

$$\leq \int_K q(t_1,x,z)|A(t_2-t_1,z)|dz + \int_{K^c} q(t_1,x,z)|A(t_2-t_1,z)|dz$$

$\leq \int_K q(t_1,x,z)|A(t_2-t_1,z)|dz + 2\int_{K^c} q(t_1,x,z)dz$, since $|A(t_2-t_1,z)| \leq 2$,

$\leq \int_K q(t_1,x,z)\, |A(t_2 - t_1,z)|dz + 2\varepsilon$, if $t_1 >> 0$.

The theorem now follows from the fact that

$$\lim_{t_2 - t_1 \to \infty} \sup_{z \in K} |A(t_2 - t_1, z)| = 0.$$

<u>WEAK ERGODIC THEOREM.</u>

$$\lim_{t \to \infty} E^{Q_x}[|\frac{1}{t} \int_0^t f(X(s))ds - f(x) \phi(x)dx| > \varepsilon] = 0.$$

<u>Proof.</u>

$$E^{Q_x}[|\frac{1}{t} \int_0^t f(X(s))ds - \int f(x) \phi(x)dx| > \varepsilon]$$

$$\leq \frac{1}{\varepsilon^2} E^{Q_x}[|\frac{1}{t} \int_0^t f(X(s))ds - \int f(y) \phi(y)dy|^2],$$

by Tchebychev's inequality. We show that the right side $\to 0$ as $t \to \infty$. Now

$$E^{Q_x} [|\frac{1}{t} \int_0^t f(X(s))ds - \int f\phi|^2]$$

$$= E^{Q_x}[|\frac{1}{t^2} \int_0^t \int_0^t f(X(\sigma_1))f(X(\sigma_2))d\sigma_1 \, d\sigma_2 + (\int f\phi dy)$$

$$- 2 \frac{1}{t} \int_0^t f(X(\sigma)) \, d\sigma \int f\phi dy]$$

Also

$$\sup_{x \in K} |E^{Q_x}[f(X(t)) - \int f(y) \phi(y)dy]|$$

$$= \sup_{x \in K} |\int q(t,x,y) f(y)dy - \int f(y)\phi(y)dy|$$

$$\leq \|f\|_\infty \sup_{x \in K} \int |q(t,x,y) - \phi(y)|dy;$$

the right hand side tends to 0 as t tends to $+\infty$. Consider

$$|E^{Q_x}(\frac{1}{t}\int_0^t f(X(\sigma))\,d\sigma - \int f(y)\,\phi(y)\,dy)|$$

$$= |E^{Q_x}(\frac{1}{t}\int_0^T f(X(\sigma))d\sigma - \int f(y)\phi(y)dy + \frac{1}{t}\int_0^t f(X(\sigma))d\sigma)\ ,0 \leq T \leq t,$$

$$\leq \frac{1}{t}|\int_0^T E^{Q_x} f(X(\sigma))d\sigma - T\int f(y)\phi(y)dy| +$$

$$+ |E^{Q_x}(\frac{1}{t}\int_T^t f(X(\sigma))d\sigma - (\frac{t-T}{t})\int f(y)\phi(y)dy)|.$$

Given $\varepsilon > 0$ choose T large so that

$$|E^{Q_x}(f(X(\sigma)) - \int f(y)\phi(y)dy| \leq \varepsilon,\ (\sigma \geq T)\ .$$

Then

$$|E^{Q_x}(\frac{1}{t}\int_0^t f(X(\sigma))d\sigma - \int f(y)\phi(y)\,dy| \leq$$

$$\leq |\frac{1}{t}\int_0^T E^{Q_x}[f(X(\sigma))] - \frac{T}{t}\int f(y)\phi(y)dy]| + \frac{t-T}{t}\varepsilon$$

$$\leq 2\varepsilon$$

provided t is large. Thus

$$\lim_{t\to+\infty} E^{Q_x}[\frac{1}{t}\int_0^t f(X(\sigma))d\sigma] = \int f\phi dy.$$

To prove the result we have therefore only to show that

$$\lim_{t\to+\infty} E^{Q_x}(\frac{1}{t^2}\int_0^t\int_0^t f(X(\sigma_1))f(X(\sigma_2))d\sigma_1 d\sigma_2) = [\int f(y)\phi(y)dy]^2$$

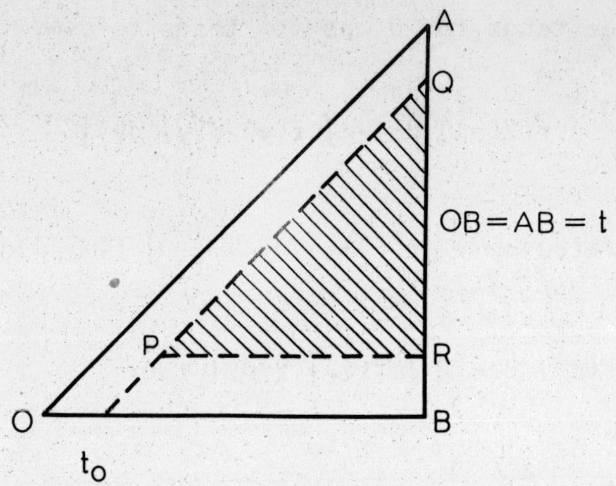

PQR is the region $\sigma_2 \geq t_0$, $\sigma_1 - \sigma_2 \geq t_0$.

Let I

$$= E^{Q_x}\left(\frac{1}{t} \int_0^t \int_0^t f(X(\sigma_1))\, f(X(\sigma_2))\, d\sigma_1\, d\sigma_2\right) - \left(\int f\phi dy\right)$$

$$= \frac{2}{t^2} \int [E^{Q_x}(f(X(\sigma_1))f(X(\sigma_2))) - (\int f(y)\phi(y)dy)^2] d\sigma_1 d\sigma_2$$

$$0 \leq \sigma_2 \leq \sigma_1 \leq t.$$

Then

$$|I| \leq \frac{2}{t^2} \int_{\Delta PQR} |E^{Q_x}(f(X(\sigma_1))f(X(\sigma_2))) - (\int f(y)\phi(y)dy)^2| d\sigma_1 d\sigma_2$$

$$+ \frac{2}{t^2} \cdot 2 \|f\|_\infty^2 \quad [\text{area of } OAB - \text{area of } PQR]$$

By the Ergodic theorem the integrand of the first term on the right can be made less than $\varepsilon/2$ provided t_0 is large (see diagram). Therefore

$$|I| \le \frac{\varepsilon}{2} \cdot \frac{2}{t^2} \text{ area of } PQR + \frac{4}{t^2} \|f\|_\infty^2 \left[\frac{t^2}{2} - \left(\frac{t-2t_0}{2}\right)^2\right]$$

$$\le \frac{\varepsilon}{2} + \frac{2\|f\|_\infty^2}{t^2}[4tt_0 - 4t_0^2].$$

$$< \varepsilon$$

if t is large. This completes the proof of the theorem.

33. APPLICATION OF STOCHASTIC INTEGRAL

LET b BE A bounded function. For every Brownian measure P_x on $\Omega = C([0,\infty); \mathbb{R}^d)$ we have a probability measure Q_x on $(\Omega, \mathcal{F})$.

Problem. Let $q(t,x,A) = Q_x(X_t \in A)$. $q(t,x,.)$ is a probability measure on $\mathbb{R}^d$. We would like to know if $q(t,x,.)$ is given by a density function on $\mathbb{R}^d$ and study its properties.

Step (i). $q(t,x,.)$ is absolutely continuous with respect to the Lebesgue measure.

For, $p(t,x,A) = P_x(X(t) \in A)$ is given by a density function. Therefore $p(t,x,.) >> m_d$ (Lebesgue measure). Since

$$Q_x << P_x \text{ on } \mathcal{F}_t,$$

$$q(t,z,.) \leq M_d \text{ on } \mathcal{F}_t.$$

Step (ii). Let $q(t,x,y) \geq 0$ be the density function of $q(t,x,.)$ and write $p(t,x,y)$ for the density of $p(t,x,.)$. Let $1 < \alpha < \infty$. Put

$$r(t,x,y) = \frac{q(t,x,y)}{p(t,x,y)}.$$

$$\int_{\mathbb{R}^d} q^\alpha dy = \int r^\alpha p^\alpha dy$$

$$= \int r^\alpha p^{1/\alpha}\, p^{\frac{\alpha-1}{\alpha}}\, dy \leq (\int r^{\alpha^2} p dy)^{1/\alpha} \times$$

$$\times (\int p^{\alpha+1}\, dy)^{\alpha-1/\alpha}$$

<u>Step (iii)</u>. $Q_x(X(t) \in A) = \int q(t,x,y)dy$

$$= \int r(t,x,y)\, p(t,x,y)dy$$

$$= \int r(t,x,y)\, P_x(X_t \in dy).$$

Therefore

$$\left.\frac{dQ_x}{dP_x}\right|_{\mathcal{F}_t^t} = r(t,x,y)$$

Therefore

$$\left(\int r^{\alpha^2} p dy\right)^{1/\alpha} = \left\| \left.\frac{dQ_x}{dP_x}\right|_{\mathcal{F}_t^t} \right\|^{\alpha}_{\alpha^2, P_x}$$

$$\leq \left\| \left.\frac{dQ_x}{dP_x}\right|_{\mathcal{F}_t} \right\|^{\alpha}_{\alpha^2, P_x}, \quad \text{since } \mathcal{F}_t^t \subset \mathcal{F}_t,$$

$$= \{E^{P_x}[Z(t)^{\alpha^2}]\}^{1/\alpha}$$

$$= \{E^{P_x}[\exp(\alpha^2 \int_0^t < b, dX > - \frac{\alpha^2}{2} \int_0^t |b|^2 ds)]\}^{1/\alpha}$$

$$= \{E^{P_x}[\exp(\alpha^2 \int_0^t < b, dX > - \frac{\alpha^4}{2} \int_0^t |b|^2 ds + \frac{\alpha^4 - \alpha^2}{2} \int_0^t |b|^2 ds)]\}^{1/\alpha},$$

i.e.,

$$\left(\int r^{\alpha^2} p dy\right)^{1/\alpha} \leq \{E^{P_x}[\exp(\frac{\alpha^4 - \alpha^2}{2} ct + \alpha^2 \int_0^t < b , dX > - \frac{\alpha^4}{2} \int_0^t |b|^2 ds)]\}^{1/\alpha}$$

where c is such that $|b|^2 \leq c$. Using Schwarz inequality we then get

$$\left(\int r^{\alpha^2} p dy\right)^{1/\alpha} \leq [\exp(\frac{\alpha^4 - \alpha^2}{2} ct)]^{1/\alpha}.$$

Hence

$$\int q^{\alpha} dy \leq \left(\exp\left[\frac{\alpha^4-\alpha^2}{2} ct\right]\right)^{1/\alpha}\left(\int p^{\alpha+1} dy\right)^{\alpha-1/\alpha}$$

Significance. Pure analytical objects like $q(t,x,y)$ can be studied using stochastic integrals.

APPENDIX

LANGUAGE OF PROBABILITY

DEFINITION.

A probability space is a measure space $(\Omega, \mathcal{B}, P)$ with $P(\Omega) = 1$. P is called a *probability measure* or simply a *probability*. Elements of $\mathcal{B}$ are called ***events***. A measurable function $X: (\Omega, \ \ ) \to R^d$ is called d -dimensional *random variable*. Given the random variable X, define $F: R^d \to R$ by

$$F((a_1, .. a_n)) = P\{w: X_i(w) < a_i, \text{ for } i = 1,2,.,d\}$$

where $X = (X_1, X_2, .., X_d)$. Then F is called the *distribution function* of the random variable X. For any random variable X, $\int X\, dP = (\int X_1 dP, .., \int X_d\, dP)$, if it exists, is called *mean* of X or *expectation* of X and is denoted by $E(X)$. Thus $E(X) = \int X dP = \mu$. $E(X^n)$, where $X^n = (X_1^n, X_2^n, .., X_d^n)$ is called the nth ***moment*** about zero. $E((X - \mu)^n)$ is called the nth *central moment*. The 2nd central moment is called ***variance*** and is denoted by σ^2 we have the following.

TCHEBYSHEV'S INEQUALITY.

Let X be a one-dimensional random variable with mean and variance μ. Then for every $\varepsilon > 0, P\{w: |X(w) - \mu| \geq \varepsilon\} \leq \sigma^2/\varepsilon^2$.

GENERALISED TCHEBYSHEV'S INEQUALITY. Let $f: R \to R$ be measurable such that $f(u) = f(-u)$, f is strictly positive and increasing on

$(0,\infty)$. Then for any random variable $X: \Omega \to R$,

$$P(w:|X(w)| > \varepsilon) \quad \frac{E(f(X))}{f(\varepsilon)}$$

for every $\varepsilon > 0$.

For any random variable $X: \Omega \to R^d$, $\phi(t) = E(e^{itX}): R^d \to C$ is called the *characteristic function* of X. Here $t = (t_1,..,t_d)$ and $= t_1X_1 + t_2X_2 + .. + t_dX_d$.

<u>INDEPENDENCE</u>. Events $E_1,..,E_n$ are called *independent* if for every $\{i,..,i_k\} \subset \{1,2,..,n\}$ we have

$$P(E_{i_1} \cap .. \cap E_{i_k}) = P(E_{i_1}) P(E_{i_2}) ... P(E_{i_k}).$$

An arbitrary collection of events $\{E_\alpha : \alpha \in I\}$ is called independent if every finite sub-collection is independent. Let $\{\mathcal{F}_\alpha : \alpha \in I\}$ be a collection of sub-σ-algebras of $\mathcal{B}$. This collection is said to be independent if for every collection $\{E_\alpha : \alpha \in I\}$, where $E_\alpha \in \mathcal{F}_\alpha$, of events is independent. A collection of random variables $\{X_\alpha : \alpha \in I\}$ is said to be independent if $\{\sigma(X_\alpha): \alpha \in I\}$ is independent where $\sigma(X_\alpha)$ is the σ-algebra generated by X_α.

<u>THEOREM</u>. *Let* $X_1,X_2,..,X_n$ *be random vairables with* $F_{X_1},..,F_{X_n}$ *as their distribution functions and let* F *be distribution function of* $X = (X_1,..,X_n)$, $\phi_{X_1},..,\phi_{X_n}$ *the characteristic functions of* $X_1,..,X_n$ *and* ϕ *that of* $X = (X_1,..,X_n)$. $X_1,..,X_n$ *are independent if and only if* $F((a_1,..,a_n)) = F_{X_1}(a_1) .. F_{X_n}(a_n)$ *for all* $a_1,..,a_n$, *iff*

$\phi((t_1,\ldots,t_n)) = \phi_{X_1}(t_1) \ldots \phi_{X_n}(t_n)$ *for all* $t_1,\ldots,t_n$.

CONDITIONING.

THEOREM. *Let* $X: (\Omega, \mathcal{B}, P) \to R^d$ *be a random variable, with* $E(X)$ *finite, i.e. if* $X = (X_1,\ldots,X_d)$, $E(X_i)$ *is finite for each* i. *Let* $\mathcal{C}$ *be a sub-* σ *-algebra of* $\mathcal{B}$. *Then there exists a random variable* $Y: (\Omega, \mathcal{C}) \to R^d$ *such that* $\int_C Y dP = \int_C X dP$ *for every* C *in* $\mathcal{C}$. *If* Z *is any random variable with the same properties then* $Y = Z$ *almost everywhere* (P).

DEFINITION. Any such Y is called the *conditional expectation of* X *with respect to* $\mathcal{C}$ and is denoted by $E(X|\mathcal{C})$.

If $X = \chi_A$, the characteristic function of A in $\mathcal{B}$, then $E(\chi_A|\mathcal{C})$ is also denoted by $P(A|\mathcal{C})$.

Properties of conditional expectation.

1. $E(1|\mathcal{C}) = 1$.
2. $E(aX + bY|\mathcal{C}) = aE(X|\mathcal{C}) + bE(Y|\mathcal{C})$ for all real numbers a,b and random variables X,Y.
3. If X is a one-dimensional random variable and $X \geq 0$, then $E(X|\mathcal{C}) \geq 0$.
4. If Y is a bounded $\mathcal{C}$ -measurable real valued random variable and X is a one-dimensional random variable, then
$$E(YX|\mathcal{C}) = YE(X|\mathcal{C}).$$
5. If $\mathcal{D} \subset \mathcal{C} \subset \mathcal{B}$ are σ -algebras, then
$$E(E(X|\mathcal{C})|\mathcal{D}) = E(X|\mathcal{D}).$$

6. $\int_\Omega |E(X|\mathcal{D})| \; d(P|\mathcal{D}) \leq \int_\Omega E(X|\mathcal{C}) \; d(P|\mathcal{C})$.

Exercise 1. Let $(\Omega, \mathcal{B}, P)$ be a probability space, $\mathcal{C}$ a sub-σ-algebra of $\mathcal{B}$. Let $X(t,.)\ Y(t,.): \Omega \to R$ be measurable with respect to $\mathcal{B}$ and $\mathcal{C}$ respectively where t ranges over the real line. Further let $E(X(t,.)|\mathcal{C}) = Y(t,.)$ for each t. If f is a simple $\mathcal{C}$-measurable function then show that

$$\int_C X(f(w),w) \; d(P|\mathcal{C}) = \int_C Y(f(w)w) \; dP$$

for every C in $\mathcal{C}$.

[Hint. Let $A_1,..,A_n$ be a $\mathcal{C}$-measurable partition such that f is constant on each A_i. Verify the equality when C is replaced by $C \cap A_i$.]

Exercise 2. Give conditions on X,Y such that exercise 1 is valid for all bounded $\mathcal{C}$-measurable functions and prove your claim.

The next lemma exhibits conditioning as a projection on a Hilbert space.

LWMMA. *Let* $(\Omega, \mathcal{B}, P)$ *be any probability space* $\mathcal{C}$ *a sub-σ-algebra of* $\mathcal{B}$. *Then*

a) $L^2(\Omega, \mathcal{C}, P)$ *is a closed subspace of* $L^2(\Omega, \mathcal{B}, P)$.

b) *If* $\pi: L^2(\Omega, \mathcal{B}, P) \to L^2(\Omega, \mathcal{C}, P)$ *is the projection, then* $\pi(f) = E(f|\mathcal{C})$.

Proof. (a) is clear, because for any $f \in L^1(\Omega, \mathcal{C}, P)$

$$\int_\Omega f \, d(P|\mathcal{C}) = \int_\Omega f \, d P$$

(use simple function $0 \leq s_1 \leq .. \leq f$, if $f \geq 0$) and $L^2(\Omega, \mathcal{C}, P)$ is complete.

(b) To prove this it is enough to verify it for characteristic functions because both π and $f \to E(f|\mathcal{C})$ are linear and continuous.

Let $A \in \mathcal{B}$, $C \in \mathcal{C}$ then $\pi(\chi_C) = \chi_C$. As π is a projection

$$\int \pi(\chi_A) \, \overline{\chi_C} \, d(P|\mathcal{B}) = \int \chi_A \, \overline{\pi(\chi_C)} \, d(P|\mathcal{B}),$$

i.e.

$$\int_C \pi(\chi_A) \, d(P|\mathcal{B}) = \int_C \chi_A \, d(P|\mathcal{B}).$$

Since $\pi(\chi_A)$ is $\mathcal{C}$-measurable,

$$\int_C \pi(\chi_A) \, d(P|\mathcal{B}) = \int_C \pi(\chi_A) \, d(P|\mathcal{C})$$

Therefore

$$\int_C \pi(\chi_A) \, d(P|\mathcal{C}) = \int_C \chi_A \, d(P|\mathcal{B}), \ \forall \, C \text{ in } \mathcal{C} .$$

Hence

$$\pi(\chi_A) = E(\chi_A | \mathcal{C}).$$

KOLMOGOROV'S THEOREM.

Statement. Let A be any nonempty set and for each finite ordered subset $(t_1, t_2, .., t_n)$ of A [i.e. $(t_1, .., t_n)$ an ordered n-tuple with t_i in A], let $P_{(t_1, .., t_n)}$ be a probability on the Borel sets in $R^{dn} = R^d \times R^d \times .. \times R^d$. Assume that the family $P_{(t_1,..,t_n)}$ satisfies the following two conditions

(i) Let $\tau: \{1,2,..,n\} \to \{1,2,..,n\}$ be any permutation and $f_\tau: R^{dn} \to R^{dn}$ be given by

$$f_\tau((x_1,..,x_n)) = (x_{\tau(1)},..,x_{\tau(n)}).$$

We have

$$P_{(t_{\tau(1)},..,t_{\tau(n)})}(E) = P_{(t_1,..,t_n)}(f_\tau^{-1}(E))$$

for every Borel set E of R^{dn}. In short, we write this condition as $P_{\tau t} = P_t \tau^{-1}$.

(ii) $$P_{(t_1,..,t_n)}(E) = P_{(t_1,t_2,..,t_n,t_{n+1},..t_{n+m})}(E \times R^{dm})$$

for all Borel sets E of R^{dn} and this is true for all $t_1,..,t_n,t_{n+1},..,t_{n+m}$ of A.

Then, there exists a probability space $(\Omega, \mathcal{B}, P)$ and a collection of random variables $\{X_t: t \in A\}: (\Omega, \mathcal{B}) \to R^d$ such that

$$P_{(t_1,..,t_n)}(E) = P\{w: (X_{t_1}(w),..,X_{t_n}(w)) \in E\}$$

for all Borel sets E of R^{dn}.

<u>Proof.</u> Let $\Omega = \Pi\{R_t^d: t \in A\}$ where $R_t^d = R^d$ for each t. Define $X_t: \Omega \to R^d$ to be the projection given by $X_t(w) = w(t)$. Let $\mathcal{B}_0$ be the algebra generated by $\{X_t: t \in A\}$ and $\mathcal{B}$ the σ-algebra generated by $\{X_t: t \in A\}$. Having got Ω and $\mathcal{B}$ we have to construct a probability P on $(\Omega, \mathcal{B})$ satisfying the conditions of the theorem.

Given $t_1,..,t_n$ define

$$\pi_{(t_1,..,t_n)} : \Omega \to R^d \times R^d \times .. \times R^d \text{ (n times)}$$

by

$$\pi_{(t_1,..,t_n)}(w) = (w(t_1),..,w(t_n)).$$

It is easy to see that every element of $\mathcal{B}_0$ is $\pi^{-1}_{(t_1,..,t_n)}(E)$ for suitable $t_1,..,t_n$ in A and a suitable Borel set E of R^{dn} Define P on $\mathcal{B}_0$ by $P(\pi^{-1}_{(t_1,..,t_n)}(E)) = P_{(t_1,..,t_n)}(E)$.

Conditions (1) and (2) ensure that P is a well-defined function on $\mathcal{B}_0$ and that, as $P_{(t_1,..,t_n)}$ are measures, P is finitely additive on $\mathcal{B}_0$.

<u>CLAIM.</u> Let $C_1 \supset C_2 \supset .. \supset C_n \supset ..$ be a decreasing sequence in $\mathcal{B}_0$ with $\lim_{n \to \infty} P(C_n) \geq \delta > 0$. Then $\bigcap C_n$ is non-empty. Once the claim is proved, by Kolmorov's theorem on extension of measures, the finitely additive set function P can be extended to a measure P on $\mathcal{B}$. One easily sees that P is a required probability measure.

<u>Proof of the Claim.</u> As $C_n \in \mathcal{B}_0$, we have

$$C_n = \pi^{-1}_{(t_1^{(n)}, \, .. \, , \, t_{k(n)}^{(n)})}(E_n) \text{ for suitable } t_i^{(n)} \text{ in A}$$

and Borel set E_n in $R^{dk(n)}$. Let $T_n = (t_1^{(n)},..,t_{k(n)}^{(n)})$ and $A_n = \{t^{(n)}, .. \, t_{k(n)}^{(n)}\}$. We can very well assume that A_n is increasing with n. Choose a compact subset E'_n of E_n such that

$$P_{T_n}(E_n - E'_n) \leq \delta/2^{n+1}.$$

If $C'_n = \pi^{-1}_{T_n}(E'_n)$, then $P(C_n - C'_n) \leq \delta/2^{n+1}$. If

$C''_n = C'_1 \cap C'_2 \cap .. \cap C'_n$ then $C''_n \subset C'_n \subset C_n$, C''_n is decreasing and

$$P(C''_n) \geq P(C_n) - \sum_{i=1}^{n} P(C_i - C'_i) \geq \delta/2.$$

We prove $\bigcap C''_n$ is not empty, which proves the claim.

Choose w_n in C''_n. As $\pi_{T_1}(w_n)$ is in the compact set E_1 for all n, choose a subsequence

$$n_1^{(1)}, n_2^{(1)}, \ldots \text{ of } 1,2,.. \text{ such that } \pi_{T_1}(\omega_{n_k}(1))$$

converges as $k \to \infty$. But for finitely many $n_k^{(1)}$'s, $\pi_{T_2}(\omega_{n_m}(1))$ is in the compact set E'_2. As before choose a subsequence $n_k^{(2)}$ of $n_k^{(1)}$ such that $\pi_{T_1}(\omega_{n_k}(2))$ converges as $k \to \infty$. By the diagonal process obtain a subsequence, w^*_n of w_n such that $\pi_{T_m}(w^*_n)$ converges as $n \to \infty$ for all m. Thus, if t is in

$$\bigcup_{m=1}^{\infty} A_m, \text{ then } \lim_{n \to \infty} w^*_n(t) = x_t$$

exists. Define w by $w(t) = 0$ if $t \in A \bigcup_{m=1}^{\infty} A_m$, $w(t) = x_t$ if $t \in \bigcup_{m=1}^{\infty} A_m$. One easily sees that $w \in \bigcap_{n=1}^{\infty} C''_n$, completing the proof of the theorem.

MARTINGALES.

DEFINITION. Let $(\Omega, \mathcal{F}, P)$ be a probability space, $(T, \leq)$ a totally ordered set. Let $(\mathcal{F}_t)_{t \in T}$ be an increasing family of sub- σ -algebras of $\mathcal{F}$. A collection $(X_t)_{t \in T}$ of random variable on Ω is called a *martingale* with respect to the family $(\mathcal{F}_t)_{t \in T}$ if

(i) $E(|X_t|) < \infty, \forall t \in T;$

(ii) X_t is $\mathcal{F}_t$- measurable for each $t \in T$;

(iii) $E(X_t | \mathcal{F}_s) = X_s$ a.s. for each s,t in T with $t \geq s$. (Markov property).

If instead of (iii) one has

(iii)' $E(X_t | \mathcal{F}_s) \geq (\leq) X_s$ a.s.,

then $(X_t)_{t \in T}$ is called a *submartingale* (respectively *supermartingale*).

From the definition it is clear that $(X_t)_{t \in T}$ is a submartingale if and only if $(-X_t)_{t \in T}$ is a supermartingale, hence it is sufficient to study the properties of only one of these.

T is usually any one of the following sets

$$[0,\infty),\ N, Z,\ \{1,2,..,n\},\ [0,\infty] \text{ or } N \cup \{\infty\}.$$

Examples.

1) Let $(X_n)_{n=1,2\,..}$ be a sequence of independent random variable with

$$E(X_n) = 0.$$

Then $Y_n = X_1 + \ldots + X_n$ is a martingale with respect to $(\mathcal{F}_n)_{n=1,2 \ldots}$

where

$$\mathcal{F}_n = \sigma\{Y_1, \ldots, Y_n\} = \sigma\{X_1, \ldots, X_n\} .$$

Proof. By definition, each Y_n is $\mathcal{F}_n$ - measurable.

$$E(Y_n) = 0.$$

$$E((X_1 + .. + X_n + X_{n+1} + .. + X_{n+m}) | \sigma\{X_1, .., X_n\})$$

$$= X_1 + .. + X_n + E((X_{n+1} + .. + X_{n+m}) | \sigma\{X_1, .., X_n\})$$

$$= Y_n + E(X_{n+1} + .. + X_{n+m}) = Y_n .$$

2) Let $(\Omega, \mathcal{F}, P)$ be a probability space, Y a random variable with $E(|Y|) < \infty$. Let $\mathcal{F}_t \subset \mathcal{F}$ be a σ -algebra such that $\forall\, t \in [0,\infty)$

$$\mathcal{F}_t \subset \mathcal{F}_s \text{ if } t \leq s.$$

If $X_t = E(Y | \mathcal{F}_t)$, X_t is a martingale with respect to $(\mathcal{F}_t)$.

Proof. (i) By definition, X_t is $\mathcal{F}_t$ -measurable.

(ii) $E(X_t) = E(Y)$ (by definition) $< \infty$.

(iii) if $t \geq s$,

$$E(X_t | \mathcal{F}_s) = E(E(Y | \mathcal{F}_t) | \mathcal{F}_s) = E(Y | \mathcal{F}_s) = X_s$$

Exercise 1. $\Omega = [0,1]$, $\mathcal{F} = \sigma$ -algebra of all Borel sub set of Ω ,

P = Lebesgue measure.

Let $\mathcal{F}_n$ = -algebra generated by the sets $[0, \frac{1}{2^n})\ [\frac{1}{2^n}, \frac{2}{2^n})$; ...

$\ldots, [\frac{2^n-1}{2^n}, 1]$. Let $f \in L'[0,1]$ and define

$$X_n(w) = 2^n \left(\sum_{j=1}^{2^n-1} \chi_{[\frac{j-1}{2^n}, \frac{j}{2^n})} \int_{j-1/2^n}^{j/2^n} f\, dy + \right.$$

$$\left. + \chi_{[\frac{2^n-1}{2^n}, 1]} \int_{2^n-1/2^n}^{1} f\, dy \right)$$

Show that (X_n) is a martingale relative to $(\mathcal{F}_n)$.

Exercise. Show that a submartingale or a supermartingale $\{X_s\}$ is a martingale iff $E(X_s) = \text{constant}$.

THEOREM. *If* $(X_t)_{t \in T}$, $(Y_t)_{t \in T}$ *are supermartingales then*

(i) $(aX_t + bY_t)_{t \in T}$ *is a supermartingale,* $\forall\, a,b \in \mathbb{R}^+ = [0,\infty)$.

(ii) $(X_t \wedge Y_t)_{t \in T}$ *is a supermartingale.*

Proof. (i) Clearly $Z_t = aX_t + bY_t$ is $\mathcal{F}_t$ -measurable and $E(|Z_t|) \le aE(|X_t|) + bE(|Y_t|) < \infty$.

$$E(aX_t + bY_t | \mathcal{F}_s) = aE(X_t|\mathcal{F}_s) + bE(Y_t|\mathcal{F}_s)$$

$$\le aX_s + bY_s = Z_s, \text{ if } t \ge s .$$

(ii) Again $X_t \wedge Y_t$ is $\mathcal{F}_t$ -measurable and $E(|X_t \wedge Y_t|) < \infty$,

$$E(X_t \wedge Y_t|\mathcal{F}_s) \le E(X_t|\mathcal{F}_s) \le X_s.$$

Similarly

$$E(X_t \wedge Y_t | \mathcal{F}_s) \leq E(Y_t | \mathcal{F}_s) \leq Y_s, \quad \text{if} \quad t \geq s.$$

Therefore

$$E(X_t \wedge Y_t | \mathcal{F}_s) \leq X_s \wedge Y_s.$$

<u>Jensen's Inequality</u>. Let X be a random variable in $(\Omega, \mathcal{B}, P)$ with $E(|X|) < \infty$ and let $\phi(x)$ be a convex function defined on the real line such that $E(|\phi_0 X|) < \infty$. Then

$$\phi(E(X|\mathcal{C})) \leq E(\phi_0 X| \mathcal{C}) \quad \text{a.e.}$$

where $\mathcal{C}$ is any sub- σ -algebra of $\mathcal{B}$.

<u>Proof</u>. The function ϕ being convex, there exist sequences $a_1, a_2, .. a_n, .., b_1, b_2, ..$ of real numbers such that $\phi(x) = \sup_n (a_n x + b_n)$ for each x. Let $L_n(x) = a_n x + b_n$. Then

$$L_n(E(X|\mathcal{C})) = E(L_n(X)|\mathcal{C}) \leq E(\phi(X)|\mathcal{C})$$

for all n so that

$$\phi(E(X|\mathcal{C})) \leq E(\phi(X)|\mathcal{C}).$$

<u>Exercise</u>. (a) If $\{X_t : t \in T\}$ is a martingale with respect to $\{\mathcal{F}_t : t \in T\}$ and ϕ is a convex function on the real line such that $E(|\phi(X_t)|) < \infty$ for every t, then $\{\phi(X_t)\}$ is a martingale.

(b) If $(X_t)_{t \in T}$ is a submartingale and $\phi(x)$ is a convex function and nondecreasing and if $E(|\phi_0 X_t|) < \infty, \forall t$ then $\{\phi(X_t)\}$ is a submartingale. (Hint: Use Jensen's inequality).

DEFINITION. Let $(\Omega, \mathcal{B}, P)$ be a probability space and $(\mathcal{F}_t)_{t \in [0,\infty)}$ an increasing family of sub- σ -algebras of $\mathcal{F}$. Let $(X_t)_{t \in [0,\infty)}$ be a family of random variables on Ω such that X_t is $\mathcal{F}_t$ -measurable for each $t \geq 0$. (X_t) i s said to be *progressively measurable* if

$$X: [0,t] \times \Omega \to \mathbb{R} \text{ defined by } X(s,w) = X_s(w)$$

is measurable with respect to the σ -algebra $\mathcal{B}[0,t] \times \mathcal{F}_t$ for every t.

Stopping times. Let us suppose we are playing a game of chance, say, tossing a coin. The two possible outcomes of a toss are H (Heads) and T (Tails). We assume that the coin is unbiased so that the probability of getting a head is the same as the probability of getting a tail. Further suppose that we gain +1 for every head and lose 1 for every tail. A game of chance of this sort has the following features.

1. A person starts playing with an initial amount N and finishes with a certain amount M.

2. Certain rules are specified which allow one to decide when to stop playing the game. For example, a person may not have sufficient money to play all the games, in which case he may decide to play only a certain number of games.

It is obvious that such a game of chance is fair in that it is neither advantageous nor disadvantageous to play such a game and on the average M will equal N, the initial amount. Furthermore, the stopping rules that are permissible have to be reasonable. The following type of stopping rule is obviously unreasonable.

Rule. If the first toss is a tail the person quits at time 0 and if the first toss is a head the person quits at time $t = 1$.

This rule is unreasonable because the decision to quit is made on the basis of a future event, whereas if the game is fair this decision should depend only on the events that have already occured. Suppose, for example, 10 games are played, then the quitting times can be 0, 1, 2, .., 10. If $\xi_1,..,\xi_{10}$ are the outcomes ($\xi_i = +1$ for H, $\xi_i = -1$ for T) then the quitting time at the 5th stage (say) should depend only on $\xi_1,..,\xi_4$ and not any of $\xi_5,..\xi_{10}$. If we denote $\xi = (\xi_1,.., \xi_{10})$ and the quitting time τ as a function of ξ then we can say that $\{\xi: \tau = 5$ depends only on $\xi_1,..,\xi_4\}$. This leads us to the notion of stopping times.

DEFINITION. Let $(\Omega, \mathcal{F}, P)$ be a probability space, $(\mathcal{F}_t)_{t \in [0,\infty)}$ an increasing family of sub- σ -algebras of $\mathcal{F}$. $\tau: \Omega \to [0,\infty]$ is called a *stopping time* or *Markov time* (or a random variable independent of the future) if

$$\{w: \tau(w) \leq t\} \in \mathcal{F}_t \quad \text{for each} \quad t \geq 0.$$

Observe that a stopping time is a measurable function with respect to $\sigma(\bigcup \mathcal{F}_t) \subset \mathcal{F}$.

Examples.

1. τ = constant is a stopping time.

2. For a Brownian motion (X_t), the hitting time of a closed set is stopping time.

Exercise 3. Let $\mathcal{F}_{t+} \underset{\text{Def}}{\equiv} \bigcap_{s>t} \mathcal{F}_s \equiv \mathcal{F}_t$.

[If this is satisfied for every $t \geq 0$, $\mathcal{F}_t$ is said to be *right continuous*]. If $\{\tau < t\} \in \mathcal{F}_t$ for each $t \geq 0$, then τ is a stopping time. (Hint: $\{\tau \leq t = \bigcap_{n=k}^{\infty} \{\tau < t + 1/n\}$ for every k).

We shall denote by $\mathcal{F}_\infty$ the σ-algebra generated by $\bigcup_{t \in T} \mathcal{F}_t$. If τ is a stopping time, we define

$$\mathcal{F}_\tau = \{A \in \mathcal{F}_\infty : A \cap \{\tau \leq t\} \in \mathcal{F}_t, \forall\, t \geq 0\}$$

Exercise 4. (a) Show that $\mathcal{F}_\tau$ is a σ-algebra. (If $A \in \mathcal{F}_\tau$, $A^c \cap \{\tau \leq t\} = \{\tau \leq t\} - A \cap \{\tau \leq t\}$).

(b) If $\tau = t$ (constant) show that $\mathcal{F}_\tau = \mathcal{F}_t$.

THEOREM. *Let* τ *and* σ *be stopping times. Then*

(i) $\tau + \sigma, \tau \vee \sigma, \tau \wedge \sigma$ *are all stopping times.*

(ii) *If* $\sigma \leq \tau$, *then* $\mathcal{F}_\sigma \subset \mathcal{F}_\tau$.

(iii) τ *is* $\mathcal{F}_\tau$*-measurable.*

(iv) *If* $A \in \mathcal{F}_\sigma$, *then* $A \cap \{\sigma = \tau\}$ *and* $A \cap \{\sigma \leq \tau\}$ *are in* $\mathcal{F}_{\sigma \wedge \tau} \subset \mathcal{F}_\sigma \cap \mathcal{F}_\tau$. *In particular,*

$\{\tau < \sigma\}, \{\tau = \sigma\}, \{\tau > \sigma\}$ *are all in* $\mathcal{F}_\tau \cap \mathcal{F}_\sigma$.

(v) *If* τ' *is* $\mathcal{F}_\tau$ *-measurable and* $\tau' \geq \tau$, *then* τ' *is a stopping time.*

(vi) *If* $\{\tau_n\}$ *is a sequence of stopping times, then* $\underline{\lim}\ \tau_n$, $\overline{\lim}\ \tau_n$ *are also stopping times provided that* $\mathcal{F}_{t+} = \mathcal{F}_t, \forall\, t \geq 0.$

(vii) *If* $\tau_n \downarrow \tau$, *then* $\mathcal{F}_\tau = \bigcap_{n=1}^{\infty} \mathcal{F}_{\tau_n}$ *provided that* $\mathcal{F}_{t+} = \mathcal{F}_t, \forall\, t \geq 0.$

Proof.

(i) $\{\sigma + \tau\} > t\}\ \{\sigma + \tau > t, \tau \leq t, \sigma \leq t\} \cup \{\tau > t\} \cup \{\sigma > t\}, \{\sigma+\tau > t, \sigma \leq t\} =$

$$= \bigcup_{\substack{r \in \mathbb{Q} \\ 0 \leq r \leq t}} \{\sigma > r > t - \tau, \tau \leq t, \sigma \leq t\}$$

($\mathbb{Q}$ = set of rationals)

$$\{\sigma > r > t - \tau, \tau \leq t, \sigma \leq t\} = \{t \geq \sigma > r\} \cap \{t \geq \tau > t - r\}$$

$$= \{\sigma \leq t\} \cap \{\sigma \leq r\}^c \cap \{\tau \leq t\} \cap \{\tau \leq t - r\}^c.$$

The right side is in $\mathcal{F}_t$. Therefore $\sigma + \tau$ is a stopping time.

$$\{\tau \vee \sigma \leq t\} = \{\tau \leq t\} \cap \{\sigma \leq t\}$$

$$\{\tau \wedge \sigma > t\} = \{\tau > t\} \cap \{\sigma > t\}$$

(ii) Follows from (iv).

(iii) $\{\tau \leq t\}\ \ \{\tau \leq s\} = \{\tau \leq t \wedge s\} \in \mathcal{F}_{t \wedge s} \subset \mathcal{F}_s, \forall\, s \geq 0.$

(iv) $A \cap \{\sigma < \tau\} \cap \{\sigma \wedge \tau \leq t\} = [A \cap \{\sigma \leq t < \tau\}]$

$$\cup [A \cap \bigcup_{\substack{r \in \mathbb{Q} \\ 0 \leq r \leq t}} \{\sigma \leq r < \tau\} \cap \{\tau \leq t\}] \in \mathcal{F}_t.$$

$$A \cap \{\sigma \le \tau\} \cap \{\sigma \wedge \tau \le t\} = A \cap \{\sigma \le \tau\} \cap \{\sigma \le t\}.$$

It is now enough to show that $(\sigma \le \tau) \in \mathcal{F}_\sigma$; but this is obvious because $(\tau < \sigma) = (\sigma \le \tau)^c$ is in $\mathcal{F}_{\sigma \wedge \tau} \subset \mathcal{F}_\sigma$. Therefore $A \cap \{\sigma \le \tau\} \in \mathcal{F}_{\sigma \wedge \tau}$ and (iv) is proved.

(v) $\{\tau' \le t\} = \{\tau' \le t\} \cap \{\tau \le t\} \in \mathcal{F}_t$ as $(\tau' \le t) \in \mathcal{F}_\tau$.

Therefore τ' is a stopping time.

(vi)
$$\underline{\lim}\ \tau_n \equiv \sup_n \inf_{k \ge n} \tau_k$$
$$= \sup_n \inf_\ell \inf \left\{\tau_n, \tau_{n+1}, \ldots, \tau_{n+\ell}\right\}.$$

By (i), $\inf \left\{\tau_n, \tau_{n+1}, \ldots, \tau_{n+\ell}\right\}$ is a stopping time. Thus we have only to prove that if $\tau_n \uparrow \tau$ or $\tau_n \downarrow \tau$ where τ_n are stopping times, then τ is a stopping time. Let $\tau_n \uparrow \tau$. Then $\{\tau \le t\} = \bigcap_{n=1}^{\infty} \{\tau_n \le t\}$ so that τ is a stopping time. Let $\tau_n \downarrow \tau$. Then

$$\{\tau \ge t\} = \bigcap_{n=1}^{\infty} \{\tau_n \ge t\}.$$

By Exercise 3, τ is a stopping time. That $\overline{\lim}\ \tau_n$ is a stopping is proved similarly.

(vii) Since $\tau \le \tau_n$, $\forall n$, $\mathcal{F}_\tau \subset \bigcap_{n=1}^{\infty} \mathcal{F}_{\tau_n}$. Let $A \in \bigcap_{n=1}^{\infty} \mathcal{F}_{\tau_n}$.

Therefore $A \cap (\tau_n < t) \in \mathcal{F}_t$, $\forall n$. $A \cap (\tau < t) = \bigcup_{m=1}^{\infty} (A \cap (\tau_m < t)) \in \mathcal{F}_t$.

Therefore $A \in \mathcal{F}_\tau$.

OPTIONAL SAMPLING THEOREM. (Discrete case). Let $\{X_1, \ldots, X_k\}$ *be a martingale relative to* $\{\mathcal{F}_1, \ldots, \mathcal{F}_k\}$. *Let* $\{\tau_1, \ldots, \tau_p\}$ *be a collection of stopping times relative to* $\{\mathcal{F}_1, \ldots, \mathcal{F}_k\}$ *such that*

$\tau_1 \leq \tau_2 \leq \dots \leq \tau_p$ a.s. *and each* τ_i *takes values in* $\{1,2,..,k\}$. *Then* $\{X_{\tau_1},.., X_{\tau_p}\}$ *is a martingale relative to* $\{\mathcal{F}_{\tau_1},..,\mathcal{F}_{\tau_p}\}$ *where for any stopping time* τ, $X_\tau(\omega) = X_{\tau(w)}(\omega)$.

Proof. It is easy to see that each X_{τ_i} is a random variable. In fact $X_{\tau_m} = \sum_{i=1}^{k} X_i \chi_{\{\tau_m = i\}}$. Let $\tau \in \{1,2,..,k\}$. Then

$$E(|X_\tau|) \leq \sum_{j=1}^{k} \int |X_j| \, dP < \infty.$$

Consider

$$(X_{\tau_j} \leq t) \cap (\tau_j \leq s) = \bigcap_{\ell \leq s} (X_\ell \leq t) \in \mathcal{F}_s.$$

Then $(X_{\tau_j} \leq t)$ is in $\mathcal{F}_{\tau_j}$, i.e. X_{τ_j} is $\mathcal{F}_{\tau_j}$ - measurable. Next we show that

(*) $$E(X_{\tau_j} \mid \mathcal{F}_{\tau_k}) \leq X_{\tau_k}, \text{ if } j \geq k.$$

(*) is true if and only if

$$\int_A X_{\tau_j} \, dP \leq \int_A X_{\tau_k} \, dP \quad \text{for every } A \in \mathcal{F}_{\tau_k}.$$

The theorem is therefore a consequence of the following

LEMMA. *Let* $\{X_1,..,X_k\}$ *be a supermartingale relative to* $\{\mathcal{F}_1,..., \mathcal{F}_k\}$. *If* τ *and* σ *are stopping times relative to* $\{\mathcal{F}_1,..., \mathcal{F}_k\}$ *taking values in* $\{1,2,..,k\}$ *such that* $\tau \leq \sigma$ *then*

$$\int_A X_\tau \, dP \geq \int_A X_\sigma \, dP \quad \textit{for every } A \in \mathcal{F}_\tau.$$

Proof. Assume first that $\sigma - \tau \leq 1$. Then

$$\int_A (X_\tau - X_\sigma)dP = \sum_{j=1}^{k} \int_{[A \cap (\tau=j) \cap (\tau < \sigma)]} (X_\tau - X_\sigma)dP$$

$$= \sum_{j=1}^{k} \int_{[A \cap (\tau=j)]} (X_j - X_{j+1})dP$$

$A \in \mathcal{F}_\tau$. Therefore $A \cap (\tau = j) \in \mathcal{F}_j$. By supermartingale property

$$\int_{[A \cap (\tau=j)]} (X_j - X_{j+1})dP \geq 0.$$

Therefore

$$\int_A (X_\tau - X_\sigma)dP \geq 0.$$

Consider now the general case $\tau \leq \sigma$. Define $\tau_n = \sigma \wedge (\tau + n)$.

Therefore $\tau_n \geq \tau$. τ_n is a stopping time taking values in $\{1,2,..,k\}$,

$$\tau_{n+1} \geq \tau_n, \tau_{n+1} - \tau_n \leq 1, \tau_k = \sigma.$$

Therefore $\int_A X_{\tau_n} dP \geq \int_A X_{\tau_{n+1}} dP, \forall A \in \mathcal{F}_{\tau_n}$. If $A \in \mathcal{F}_\tau$ then

$A \in \mathcal{F}_{\tau_n}, \forall n$. Therefore

$$\int_A X_{\tau_1} dP \geq \int_A X_{\tau_2} dP \geq .. .. \geq \int_A X_{\tau_k} dP, \forall A \in \mathcal{F}_\tau.$$

'Now $\tau_1 - \tau \leq 1$. $\tau \leq \tau_1$. Therefore

$$\int_A X_\tau dP \geq \int_A X_{\tau_1} dP \geq \int_A X_\sigma dP.$$

This completes the proof.

N.B. The equality in (*) follows by applying the argument to $\{-X_1, \ldots, -X_k\}$.

COROLLORY.1. *Let* $\{X_1, X_2, \ldots, X_k\}$ *be a super-martingale relative to* $\{\mathcal{F}_1, \ldots, \mathcal{F}_k\}$ *If* τ *is any stopping time, then*

$$E(X_k) \le E(X_\tau) \le E(X_1).$$

Proof. Follows from the fact that $\{X_1, X\ , X_k\}$ is a super-martingale relative to $\{\mathcal{F}_1,\ \ , \mathcal{F}_k\}$.

COROLLORY 2. *If* $\{X_1, X_2, \ldots, X_k\}$ *is a super-martingale relative to* $\{\mathcal{F}_1, \ldots, \mathcal{F}_k\}$ *and* τ *is any stopping time, then*

$$E(X_\tau) \le E(|X_1|) + 2E(X_k^-) \le 3 \sup_{1 \le n \le k} E(|X_n|)$$

where for any real x, $x^- = \dfrac{|x| - x}{2}$.

Proof. $X_k^- = \dfrac{|X_k| - X_k}{2}$, so $2E(X_k^-) = E(|X_k|) - E(X_k)$.

By theorem $\{X_\tau \wedge 0, X_k \wedge 0\}$ is a super-martingale relative to $\{\mathcal{F}_\tau, \mathcal{F}_k\}$. Therefore $E(X_k \wedge 0 | \mathcal{F}_\tau) \le E(X_\tau \wedge 0)$. Hence

$$E(X_k^-) \ge E(X_\tau^-) = \frac{E(|X_\tau|) - E(X_\tau)}{2}.$$

Therefore

$$E(|X_\tau|) \le 2E(X_k^-) + E(X_\tau)$$

$$\le 2E(X_k^-) + E(X_1) \le 3 \sup_{1 \le n \le k} E(|X_n|).$$

THEOREM. *Let* $(\Omega, \mathcal{F}, P)$ *be a probability space and* $(\mathcal{F}_t)_{t \ge 0}$ *an increasing family of sub-* σ *-algebras of* $\mathcal{F}$. *Let* τ *be a finite* *stopping time, and* $(X_t)_{t \geq 0}$ *a progressively measurable family (i.e.* $X: [0,\infty) \times \Omega \to \mathbb{R}$ *defined by* $X(t,w) = X_t(w)$ *is progressively measurable). If* $X_\tau(w) = X_{\tau(w)}(w)$, *then* X_τ *is* $\mathcal{F}_\tau$ *-measurable.*

Proof. We show that $\{w; X(\tau(w),w) \leq t, \tau(w) \leq s\} \in \mathcal{F}_t$ for every t. Let $\Omega_s = \{w: \tau(w) \leq s\}$; $\Omega_s \in \mathcal{F}_s$ and hence the σ-algebra induced by $\mathcal{F}_s$ on Ω_s is precisely

$$\{A \cap \Omega_s: A \in \mathcal{F}_s\} = \{A \in \mathcal{F}_s: A \subset \Omega_s\}.$$

Since $\tau(w)$ is measurable,

$$w \to (\tau(w),w) \quad \text{of} \quad \Omega_s \to [0,s] \times \Omega_s$$

is $(\mathcal{F}_s, \mathcal{B}[0,s] \times \mathcal{F}_s)$ -measurable. Since X is progressively measurable,

$$[0,s] \times \Omega_s \overset{X}{\to} \mathbb{R} \quad \text{is measurable.}$$

Therefore $\{w:X(\tau(w),w) \leq t,\tau(w) \leq s\} \in \sigma$ -algebra on Ω_s.

Therefore X_τ is $\mathcal{F}_\tau$ measurable.

The next theorem gives a condition under which $(X_t)_{t \geq 0}$ is progressively measurable.

THEOREM. *If* X_t *is right continuous in* $t, \forall w$ *and* X_t *is* $\mathcal{F}_t$ *-measurable,* $\forall t \geq 0$ *then* $(X_t)_{t \geq 0}$ *is progressively measurable.*

Proof. Define

$$X_n(t,w) = X\left(\frac{[nt] + 1}{n}, w\right). \quad \frac{[nt] + 1}{n} \downarrow t.$$

Then

$$\underset{n \to \infty}{\text{Lt}} X_n(t,w) = X(t,w) \quad \text{(by right continuity)}$$

STEP 1. Suppose T is rational, $T = m/n$ where $m \geq 0$ is an integer. Then

$$\{(t,w): 0 \leq t < T,\ X_n(t,w) \leq \alpha\}$$

$$= \bigcup_{0 \leq i \leq m-1} \left\{[\tfrac{i}{n}, \tfrac{i+1}{n}) \times X^{-1}_{\frac{i+1}{n}}(-\infty,\alpha]\right\}$$

Thus if $T = m/n$, $X_n|_{[0,T) \times \Omega}$ is $\mathcal{B}[0,T] \times \mathcal{F}_T$ -measurable. Now $T = \frac{km}{kn}$. Letting $k \to \infty$, by right continuity of $X(t)$ one gets $X|_{[0,T) \times \Omega}$ is $[0,T] \times \mathcal{F}_T$ -measurable. As $X(T)$ is $\mathcal{F}_T$ -measurable, one gets $X|_{[0,T] \times \Omega}$ is $[0,T] \times \mathcal{F}_T$ -measurable.

Step 2. Let T be irrational. Choose a sequence of rationals S_n increasing to T.

$$\{(t,w): 0 \leq t \leq T,\ X(t,w) \leq \alpha\}$$

$$= \bigcup_{n=1}^{\infty} \{(t,w): 0 \leq t \leq S_n, X(t,w) \leq \alpha\} \cup \{T\} \times X_T^{-1}(-\infty,\alpha]$$

The countable union is in $\mathcal{B}[0,T] \times \mathcal{F}_T$ by Step 1. The second member is also in $\mathcal{B}[0,T] \times \mathcal{F}_T$ as $X(T)$ is $\mathcal{F}_T$ -measurable. Thus $X|_{[0,T] \times \Omega}$ is $\mathcal{B}[0,T] \times \mathcal{F}_T$ -measurable w hen T is irrational also.

REMARK. The technique used above is similar to the one used for proving that a right continuous function $f: R \to R$ is Borel measurable.

THEOREM. *Let* $\{X_1,..,X_k\}$ *be a supermartingale and* $\lambda \geq 0$. *Then*

(1)
$$\lambda P(\sup_{1\leq n\leq k} X_n \geq \lambda) \leq E(X_1) - \int_{\left\{\sup_{1\leq n\leq k} X_n < \lambda\right\}} X_k dP$$
$$\leq E(X_1) + E(X_k^-).$$

(2)
$$\lambda P(\inf_{1\leq n\leq k} X_n \leq -\lambda) \leq - \int_{\{\inf X_n \leq -\lambda\}} X_k dP$$
$$\leq E(X_k^-).$$

Proof. Define $\tau(w) = \inf\{n: X_n \geq \lambda\}$ if $\sup_n X_n \geq \lambda$,

$$= k, \text{ if } \sup_n X_n < \lambda.$$

Clearly $\tau \geq 0$ and τ is a stopping time. If $\tau < k$, then $X_\tau(w) \geq \lambda$ for each w.

$$E(X_\tau) = \int_{(\sup X_n \geq \lambda)} X_\tau dP + \int_{(\sup X_n < \lambda)} X_\tau dP$$

$$\geq \lambda P(\sup X_n \geq \lambda) + \int_{(\sup X_n < \lambda)} X_k dP.$$

Therefore

$$E(X_1) \geq \lambda P(\sup X_n \geq \lambda) + \int_{(\sup X_n < \lambda)} X_k dP,$$

$$\lambda P(\sup X_n \geq \lambda) \leq E(X_1) - \int_{(\sup X_n < \lambda)} X_k dP \leq E(X_1)+E(X_k^-)$$

The proof of (2) is similar if we define

$$\tau(w) = \begin{cases} \inf\{n: X_n \leq -\lambda\}, & \text{if } \inf X_n \leq -\lambda, \\ k, & \text{if } \inf X_n > -\lambda. \end{cases}$$

KOLMOGOROV'S INEQUALITY (Discrete Case). *Let* $\{X_1, .., X_k\}$ *be a finite sequence of independent random variables with mean* 0. *Then*

$$P(\sup_{1 \leq n \leq k} (|X_1 +..+ X_n| \geq \lambda) \leq \frac{1}{\lambda^2} E((X_1 + X_2 +..+ X_k)^2)$$

Proof. If $S_n = X_1 +..+ X_n$, $n = 1,2,..,k$, then $\{S_1,..,S_k\}$ is a martingale with respect to $\{\mathcal{F}_1,.., \mathcal{F}_k\}$ where $\mathcal{F}_n = \sigma\{X_1,..,X_n\}$. Therefore $S_1^2,..., S_k^2$ is a submartingale (since $x \to x^2$ is convex). By the previous theorem,

$$\lambda^2 P\{\inf - S_n^2 \leq - \lambda^2\} \leq E((-S_k^2)^-)$$

Therefore

$$P\{\sup |S_n| \geq \lambda\} \leq \frac{E((-S_k^2)^-)}{\lambda^2} = \frac{E(S_k^2)}{\lambda^2}$$

$$= \frac{1}{\lambda^2} E((X_1 + X_2 +...+ X_k)^2).$$

KOLMOGOROV'S INEQUALITY (Continuous case). *Let* $\{X(t): t \geq 0\}$ *be a continuous martingale with* $E(X(0)) = 0$. *If* $0 < T < \infty$, *then for any* $\varepsilon > 0$

$$P\{w: \sup_{0 \leq s \leq T} |X(s,w)| \geq \varepsilon) \leq \frac{1}{\varepsilon^2} E((X(T))^2).$$

Proof. For any positive integer k define $Y_0 = X(0)$,

$$Y_1 = X(\frac{T}{2^k}) - X(0),\ Y_2 = X(\frac{2T}{2^k}) - X(\frac{T}{2^k}),...,Y_{2^k} = X(\frac{2^k T}{2^k}) - X(\frac{(2^k-1)}{2^k} T).$$

By Kolmogorov inequality for the discrete case, for any $\delta > 0$.

$$P(\sup_{0 \leq n \leq 2^k} |X(\frac{nT}{2^k})| > \delta) \leq \frac{1}{\delta^2} E((X(T))^2).$$

By continuity of $X(t)$, $A_k = \{w: \sup_{0 \le n \le 2^k} |X(\frac{nT}{2^k})| > \delta\}$ increases to $\{\sup_{0 \le s \le T} |X(s)| > \delta\}$ so that one gets

$$P(\sup_{0 \le s \le T} |X(s)| > \delta) \le \frac{1}{\delta^2} E((X(T))^2). \tag{1}$$

Now

$$\begin{aligned} P(\sup_{0 \le s \le T} |X(s)| \ge \varepsilon) &\le \lim_{m \to \infty} P(\sup_{0 \le s \le T} |X(s)| > \varepsilon - \frac{1}{m}) \\ &\le \lim_{m \to \infty} \frac{1}{(\varepsilon - 1/m)^2} E((X(T))^2), \text{ by (1).} \\ &= 1/\varepsilon^2 \, E((X(T))^2). \end{aligned}$$

This completes the proof.

<u>OPTIONAL SAMPLING THEOREM</u> (Countable case). *Let* $\{X_n : n \ge 1\}$ *be a supermartingale relative to* $\{\mathcal{F}_n : n \ge 1\}$. *Assume that for some* $X_\infty \in L^1$, $X_n \ge E(X_\infty | \mathcal{F}_n)$. *Let* σ, τ *be stopping times taking values in* $N \cup \{\infty\}$, *with* $\sigma \le \tau$. *Define* $X_\tau = X_\infty$ *on* $\{\sigma = \infty\}$ *and* $X_\tau = X_\infty$ *on* $\{\sigma = \infty\}$. *Then* $E(X_\tau | \mathcal{F}_\sigma) \le X_\sigma$.

<u>Proof</u>. We prove the theorem in three steps.

<u>Step 1</u>. Let $X_\infty = 0$ so that $X_n \ge 0$. Let $\tau_k = \tau \wedge k$, $\sigma_k = \tau \wedge k$. By optional sampling theorem for discrete case $E(X_{\tau_k}) \le E(X_k) \le E(X_1)$. By Fatou's lemma, $E(X_\tau) < \infty$. Again by optional sampling theorem for the discrete case,

$$E(X_{\tau_k} | \mathcal{F}_{\sigma_k}) \le X_{\sigma_k} \quad \ldots., \quad (0).$$

Let $A \in \mathcal{F}_\sigma$. Then $A \cap \{\sigma \le k\} \in \mathcal{F}_{\sigma_k}$, and by (0)

$$\int_{A \cap \{\tau \le k\}} X_\tau \, dP \le \int_{A \cap (\tau \le k)} X_{\tau_k} \, dP \le \int_{A \cap (\sigma \le k)} X_{\sigma_k} \, dP \le \int_{A \cap (\sigma \le k)} X_\sigma \, dP.$$

Letting $k \to \infty$,

$$\int_{A \cap (\tau \ne \infty)} X_\tau \, dP \le \int_{A \cap (\sigma \ne \infty)} X_\sigma \, dP. \tag{1}$$

Clearly

$$\int_{A \cap (\tau = \infty)} X_\tau \, dP = \int_A X_\infty \, dP = \int_{A \cap (\sigma = \infty)} X_\sigma \, dP \tag{2}$$

By (1) and (2), $\int_A X \, dP \le \int_A X_\sigma \, dP$, proving that

$$E(X_\tau | \mathcal{F}_\sigma) \le X_\sigma .$$

<u>Step 2.</u> Suppose $X_n = E(X_\infty | \mathcal{F}_n)$. In this case we show that $X_\tau = E(X_\infty | \mathcal{F}_\tau)$ for every stopping time so that $E(X_\tau | \mathcal{F}_\sigma) = X_\sigma$.
If $A \in \mathcal{F}_\tau$, then

$$\int_{(\tau \le k)} X_\tau \, dP = \int_{A \cap (\tau \le k)} X_\infty \, dP \quad \text{for every } k.$$

Letting $k \to \infty$,

$$\int_{A \cap (\tau \ne \infty)} X_\tau \, dP = \int_{A \cap (\tau \ne \infty)} X_\infty \, dP , \tag{1}$$

$$\int_{A \cap (\tau = \infty)} X_\tau \, dP = \int_A X_\infty \, dP = \int_{A \cap (\tau = \infty)} X_\infty \, dP \tag{2}$$

The assertion follows from (1) and (2).

Step 3. Let X_n be general. Then

$$X_n = X_n - E(X_\infty | \mathcal{F}_n) + E(X_\infty | \mathcal{F}_n).$$

Apply Step (1) to $Y_n = X_n - E(X_\infty | \mathcal{F}_n)$ and Step (2) to $Z_n = E(X_\infty | \mathcal{F}_n)$ to complete the proof.

UNIFORM INTEGRABILITY.

DEFINITION. Let $(\Omega, \mathcal{B}, P)$ be any probability space, $L^1 = L^1(\Omega, \mathcal{B}, P)$. A family $H \subset L^1$ is called *uniformly integrable* if for every $\varepsilon > 0$ there exists a $\delta > 0$ such that $\int_{(|X| \geq \delta)} |X| dP < \varepsilon$ for all X in H.

NOTE. Every uniformly integrable family is a bounded family.

PROPOSITION. *Let* X_n *be a sequence in* L^1 *and let* $X_n \to X$ *a.e. Then* $X_n \to X$ *in* L^1 *iff* $\{X_n : n \geq 1\}$ *is uniformly integrable.*

Proof. is left as an exercise.

As $\{X_n : n \geq 1\}$ is a bounded family, by Fatou's lemma $X \in L^1$. Let $\varepsilon > 0$ be given. By Egoroff's theorem there exists a set F such that $P(F) < \varepsilon$ and $X_n \to X$ uniformly on F.

$$\int |X_n - X| dP \leq \|X_n - X\|_{\infty, \Omega - F} + \int_F |X_n - X| dP$$

$$\leq \|X_n - X\|_{\infty, \Omega - F} + \int_F |X_n| dP + \int_F |X| dP$$

$$\leq \|X_n - X\|_{\infty, \Omega - F} + \int_{F \cap (|X_n| \geq \delta)} |X_n| dP + \int_{F \cap (|X| \geq \delta)} |X| dP +$$

$$+ \int_{F \cap \{|X_n| \leq \delta\}} |X_n| dP + \int_{F \cap (|X| \leq \delta)} X \, dP$$

$$\leq \|X_n - X\|_{\infty, \Omega - F} + \int_{(|X_n| \geq \delta)} |X_n| dP + \int_{(|X| \geq \delta)} |X| dP + 2\delta\varepsilon$$

The result follows by uniform integrability of $\{X, X_n : n \geq 1\}$.

COROLLORY. *Let* $\mathcal{C}$ *be any sub-* σ *-algebra of* $\mathcal{B}$. *If* $X_n \to X$ *a.e. and* X_n *is uniformly integrable, then* $E(X_n|\mathcal{C}) \to E(X|\mathcal{C})$ *in* $L^1(\Omega, \mathcal{C}, P)$.

PROPOSITION. *Let* $H \subset L^1$. *Suppose there exists an increasing convex function* $G: [0,\infty) \to [0,\infty)$ *such that* $\lim_{t \to \infty} \frac{G(t)}{t} = \infty$ *and* $\sup_{X \in H} E(G(|X|)) < \infty$. *Then the family* H *is uniformly integrable.*

Example. $G(t) = t^2$ is a function satisfying the conditions of the theorem.

Proof. (of the proposition). Let

$$M = \sup_{X \in H} E(G(|X|)).$$

Let $\varepsilon > 0$ be given. Choose $\delta > 0$ such that

$$\frac{G(t)}{t} \geq \frac{M}{\varepsilon} \quad \text{for} \quad t \geq \delta.$$

Then for X in H

$$\int_{(|X| \geq \delta)} |X| dP \leq \frac{\varepsilon}{M} \int_{(|X| \geq \delta)} G(|X|) dP \leq \frac{\varepsilon}{M} \int_\Omega G(|X|) dP \leq \varepsilon$$

REMARK. The converse of the theorem is also true.

Exercise. Let H be a bounded set in L^∞, i.e. there exists a constant M such that $\|X\|_\infty \leq M$ for all X in H. Then H is uniformly integrable.

UP CROSSINGS AND DOWN CROSSINGS.

DEFINITION. Let $a < b$ be real numbers; let $s_1, s_2, \ldots, s_k$ be also given reals. Define $i_1, i_2, \ldots, i_k$ as follows.

$$i_1 = \begin{cases} \inf \{n: s_n < a\}, \\ k, \text{ if no } s_i < a \, ; \end{cases}$$

$$i_2 = \begin{cases} \inf \{n > i_1: s_n > b\}, \\ k, \text{ if } s_n \leq b \text{ for each } n > i_1 \, ; \end{cases}$$

$$i_3 = \begin{cases} \inf \{n > i_2: s_n < a\}, \\ k, \text{ if } s_n \geq a \text{ for each } n > i_2; \end{cases}$$

and so on

Let $t_1 = s_{i_1}$, $t_2 = s_{i_2}$, ... If (t_1, t_2), (t_3, t_4), ...,
(t_{2p-1}, t_{2p}) are the only non-empty intervals and (t_{2p+1}, t_{2p+2}),.. are all empty, then p is called the of the sequence $s_1, \ldots, s_k$ for the interval $[a,b]$ and is denoted by $U(s_1, \ldots, s_k; [a,b])$.

NOTE. U (the up crossing) always takes values in $\{0,1,2,3,\ldots\}$.

DEFINITION. For any subset S of reals define

$$U(S;[a,b]) = \sup \{U(F;[a,b]): F \text{ is a finite subset of } S\}$$

The number of down crossings is defined by

$$D(S; [a,b]) = U(-S; [-b, -a]).$$

For any real valued function f on any set S we define

$$U(f, S, [a,b]) = U(f(S),[a,b]).$$

If the domain of S is known, we usually suppress it.

PROPOSITION. *Let* $a_1, a_2, \dots$ *be any sequence of real numbers and* $S = \{a_1, a_2, \dots\}$. *If* $U(S,[a,b]) < \infty$ *for all* $a < b$, *then thee sequence* $\{a_n\}$ *is a convergent sequence.*

Proof. It is clear that if $T \subset S$ then $U(T,[a,b]) \leq U(S,[a,b])$. If the sequence were not convergent, then we can find a and b such that $\liminf a_n < a < b < \limsup a_n$. Choose $n_1 < n_2 < n_3 \dots$; $m_1 < m_2 < \dots$ such that $a_{n_i} < a$ and $a_{m_i} > b$ for all i. If $T = \{a_{n_1}, a_{m_1}, a_{n_2}, a_{m_2}, \dots\}$, then $U(S; [a,b]) \geq U(T; [a,b]) = \infty$; a contradiction.

REMARK. The converse of the proposition is also true.

THEOREM. *(Doob's inequalities for up cross ings and down crossings))* *Let* $\{X_1, \dots, X_k\}$ *be a submartingale relative to* $\{\ _1, \dots, \ _k\}$ $a < b$. *Define* $U(w,[a,b]) = U(X_1(w), \dots, X_k(w); [a.b])$ *and similarly define* $D(w,[a,b])$. *Then*

(i) U, D *are measurable functions;*

(ii) $$E(U(\cdot,[a,b])) \leq \frac{E((X_k - a)+) - E((X_1 - a)^+)}{b - a};$$

(iii) $$E(D(\cdot,[a.b])) \leq E((X_k - b)^+) / (b - a).$$

Proof. (i) is left as an exercise.

(ii) Define $Y_n = (X_n - a)^+$; these are submartingales. Then clearly $Y_n \leq 0$ if and only if $X_n \leq a$ and $Y_n \geq b - a$ iff $X_n \geq b$, so that

$UY_1(w), ., Y_k(w);[0,b-a]) = U(X_1(w),..,Y_k(w); [a,b])$

Define

$$\tau_1 = 1$$

$$\tau_2 = \begin{cases} \inf \{n: Y_n = 0\} \\ k, \text{ if each } Y_n = 0 \end{cases}$$

$$\tau_3 = \begin{cases} \inf \{n > \tau_2: Y_n > b - a, \\ k, \text{ if } Y_n < b - a \text{ for each } n > \tau_2; \end{cases}$$

$$\tau_{k+1} = k.$$

As $\{Y_1, .., Y_k\}$ is a submartingale, by optional sampling theorem $Y_{\tau_1}, .., Y_{\tau_{k+1}}$ is also a submartingale. Thus

$$E(Y_{\tau_2} - Y_{\tau_1}) + E(Y_{\tau_4} - Y_{\tau_3}) + \ldots \geq 0. \qquad (1)$$

Clearly $[(Y_{\tau_3} - Y_{\tau_2}) + (Y_{\tau_5} - Y_{\tau_4}) + \ldots](w) \geq (b-a)U(Y_1(w),..Y_k(w);$

$$[0, b-a]) = (b - a) U(w, [a,b]).$$

Therefore

$$E(Y_{\tau_3} - Y_{\tau_2}) + E(Y_{\tau_5} - Y_{\tau_4})+ \ldots \geq (b-a) E(U(.,[a,b])). \ (2)$$

By (1) and (2),

$$E(Y_k - Y_1) \geq (b-a) E(U(.,[a,b]))$$

giving the result.

(iii) Let $Y_n = (X_n - a)^+$ so that

$$D(Y_1(w),..Y_k(w); [0,b-a]) = D(X_1(w),..,X_k(w); [a,b])$$

Define

$$\tau_1 = 1;$$

$$\tau_2 = \begin{cases} \inf \{n: Y_n \geq b - a\}, \\ k, \text{ if each } Y_n < b - a; \end{cases}$$

$$\tau_3 = \begin{cases} \inf \{n > \tau_2: Y_n = 0\}, \\ k, \text{ if each } Y_n > 0 \text{ for each } n > \tau_2; \end{cases}$$

$$\tau_{k+1} = k.$$

By optional sampling theorem we get

$$0 \geq E(Y_{\tau_2} - Y_{\tau_3}) + E(Y_{\tau_4} - Y_{\tau_5}) + \ldots .$$

Therefore

$$0 \geq (b-a)E(D(Y_1,\ldots,Y_k;[0,b-a]))+E((b-a)-Y_k).$$

Hence

$$E(D(.,[a,b])) \leq E((X_k - a)^+ - (b-a)) / (b-a)$$

$$\leq \frac{E((X_k-b)^+)}{(b-a)}, \text{ for } (c-a)^+-(b-a) \leq (c-b)^+ \text{ for all } c.$$

<u>COROLLORY</u>. *Let* $\{X_1,\ldots,X_k\}$ *be a supermartingale.* U,D *as in theorem. Then*

(i) $$E(D(., [a,b])) \leq \frac{E(X_1 \wedge b) - E(X_k \wedge b)}{b - a}.$$

(ii) $$E(U(., [a,b])) \leq \frac{E((X_k - b)^-)}{b - a}.$$

<u>Proof</u> (i). $E(D(.,[a.b])) = E(U(-X_1(w),\ldots, -X_k(w), [-b, -a])$

$$\leq \frac{E((-X_k + b)^+ - (-X_1 + b)^+)}{b - a}, \text{ by above theorem,}$$

$$\leq \frac{E((b \wedge X_k) - (b \wedge X_1))}{b - a}, \text{ since for}$$

since for all a,b,c, $(b-c)^+ - (b-a)^+ \leq (b \wedge a) - (b \wedge c)$.

(ii) $E(U(., [a,b])) =$

$$= E(D(-X_1(w), .., -X_k(w); [-b, -a]))$$

$$\leq \frac{E((-X_k + a)^+)}{b - a}, \text{ by theorem,}$$

$$\leq \frac{E((X_k - b)^-)}{b - a},$$

(since $(-X_k + a)^+ \leq (X_k - b)^-$,

THEOREM. *Let* $\{X_n: n = 1,2, ..\}$ *be a supermartingale relative to* $\{\mathcal{F}_n: n = 1,2, ..\}$. *Let* $(\Omega, \mathcal{F}, P)$ *be complete.*

(i) *if* $\sup_n E(X_n^-) < \infty$, *then* X_n *converges a.e. to a random variable denoted by* X_∞.

(ii) *if* $\{X_n: n \geq 1\}$ *is uniformly integrable, then also* X_∞ *exists. Further,* $\{X_n: n = 1,2, .., n = \infty\}$ *is a supermartingale with the natural order.*

(iii) *if* $\{X_n: n \geq 1\}$ *is a martingale, then* $\{X_n: n \geq 1, n = \infty\}$ *is a martingale.*

Proof. (i) Let $U(w [a,b]) = U(X_1(w), X_2(w), ..; [a,b])$. By the corollory to Doob's inequalities theorem,

$$E(U(.,[a,b]) \leq \sup_n E((X_n - b)^-) \quad \infty$$

for all $a < b$. Allowing a,b to vary over the rationals alone we find that the sequence X_n is convergent a.e.

(ii) $\sup_n E(X_n^-) \leq \sup_n E(|X_n|) < \infty$ so that X_∞ exists.

As $X_n \to X_\infty$ in L^1 we get that $\{X_n: n \geq 1, n = \infty\}$ is a supermartingale.

(iii) follows from (ii).

PROPOSITION. *Let* $\{X_t: t \geq 0\}$ *be a supermartingale relative to* $\{\mathcal{F}_t: t \geq 0\}$. $I = [r,s]$, $a < b$ *and* S *any countable dense subset. Let* $U(w, S \cap I, [a,b]) = U(., \{X_t(w): t \in S \cap I\}, [a,b])$. *Then*

$$E(U(., S \cap I, [a,b)) \leq \frac{E((X_s - b)^-)}{b - a}.$$

Proof. Let $S \cap I$ be an increasing union of finite sets F_n; then

$$E(U(.,F_n,[a,b])) \leq \frac{E((X_{\max F_n} - b)^-)}{b - a} \leq \frac{E((X_s - b)^-)}{b - a}.$$

The result follows by Fatou's lemma.

Exercise. If further X_t is continuous i.e. $t \to X_t(w)$ is continuous for each w, then prove that

$$E(U(., I, [a,b])) \leq \frac{E((X_s - b)^-)}{b - a}$$

THEOREM. *Let* $(\Omega, \mathcal{F}, P)$ *be complete and* $\{X_t: t \geq 0\}$ *a continuous supermartingale.*

(i) *If* $\sup_{t \geq 0} E(X_t^-) < \infty$, *then* X_t *converges a.e. to a random variable* X_∞.

(ii) *If* $\{X_t: t \geq 0\}$ *is uniformly integrable then also* X_∞ *exists and* $\{X_t: t \geq 0, t = \infty\}$ *is a supermartingale.*

Proof. (i) $E(U(.,[0,n],[a,b])) \leq E((X_n - b)^-) / (b - a)$

so that

$$\lim_{n \to \infty} E(U(.,[0,n],[a,b])) \leq \sup_{0 \leq s} \frac{E((X_s - b)^-)}{b - a}$$

for all $a < b$. Thus $\{X_t(w): t > 0\}$ converges a.e. whose limit is denoted by X_∞ which is measurable.

(ii) As $E(X_t^-) \leq E(|X_t|)$ by (i) X_∞ exists, the other assertion is a consequence of uniform integrability.

COROLLORY. *Let* $\{X_t: t \geq 0\}$ *be a continuous uniformly integrable. martingale. Then* $\{X_t: 0 \leq t \leq \infty\}$ *is also a martingale.*

Exercise. Let $\{X_t: t \geq 0\}$ be a continuous martingale such that for some Y with $0 \leq Y \leq 1$ $E(Y|\mathcal{F}_t) = X_t$ show that $X_t \to Y$ a.e.

LEMMA. *Let* $(\Omega, \mathcal{F}, P)$ *be a probability space,* $\mathcal{F}_1 \supset \mathcal{F}_2 \supset \mathcal{F}_3 \cdots$ *be sub-* σ *-algebras. Let* $X_1, X_2, \ldots$ *be a real valued functions measurable with respect to* $\mathcal{F}_1, \ldots, \mathcal{F}_n, \ldots$ *respectively. Let*

(i) $E(X_{n-1} | \mathcal{F}_n) \leq X_n$

(ii) $\sup_n E(X_n) < \infty$.

Then $\{X_n : n \geq 1\}$ *is uniformly integrable.*

Proof. By (i) $E(X_n)$ is increasing. By (ii) given $\varepsilon > 0$, we can find n_0 such that if $n \geq n_0$ then $E(X_n) \leq E(X_{n_0}) + \varepsilon$. For any $\delta > 0$,

$$n \geq n_0 \quad \int_{(|X_n| \geq \delta)} |X_n| dP$$

$$= E(X_n) + \int_{(X_n \leq -\delta)} -X_n dP - \int_{(X_n < \delta)} X_n dP$$

$$\leq \int_{(X_n \leq -\delta)} -X_{n_0} dP - \int_{(X_n < \delta)} X_{n_0} dP + E(X_n) \quad \text{by (i)}$$

$$\leq \varepsilon + \int_{(X_n \geq \delta)} X_{n_0} dP - \int_{(X_n \leq -\delta)} X_{n_0} dP \text{ (because } E(X_n) \leq E(X_{n_0}) + \varepsilon)$$

$$\leq \varepsilon + \int_{(|X_n| \geq \delta)} |X_{n_0}| dP$$

Thus to show uniform integrability we have only to show $P(|X_n| \geq \delta) \to 0$ uniformly in n as $\delta \to \infty$. Now

$$E(|X_n|) = E(X_n + 2X_n^-)$$
$$\leq E(X_n) + 2E(|X_1|) \quad \text{by (i)}$$
$$\leq M < \infty \quad \text{for all } n \text{ by (ii)}$$

The result follows as $P(|X_n| \geq \delta) \leq M/\delta$.

OPTIONAL SAMPLING THEOREM. (Continuous case).

Let $\{X_t : t \geq 0\}$ *be a right continuous supermartingale relative to* $\{\mathcal{F}_t : t \geq 0\}$. *Assume there exists an* $X_\infty \in L'(\Omega, \mathcal{F}, P)$ *such that* $X_t \geq E(X_\infty | \mathcal{F}_t)$ *for* $t \geq 0$ *For any stopping time* τ

taking values in $[0,\infty]$, *let* $X_\tau = X_\infty$ *on* $\{\tau = \infty\}$. *Then*

(i) X_τ *is integrable.*

(ii) *If* $\sigma \leq \tau$ *are stopping times, then*

$$E(X_\tau | \quad _\sigma) \leq X_\sigma .$$

Proof. Define

$$\sigma_n = \frac{[2^n\sigma] + 1}{2^n} , \tau_n = \frac{[2^n\sigma] + 1}{2^n} .$$

Then σ_n, τ_n are stopping times, $\sigma_n \leq \tau_n, \sigma \leq \sigma_n, \tau \leq \tau_n$.

σ_n, τ_n take values in $D_n = \{\infty, 0, 1/2^n, 2/2^n, .., 1/2^n, ..\}$ so that we have $E(X_{\tau_n} | \mathcal{F}_{\sigma_n}) \leq X_{\sigma_n}$. Thus if, $A \in \mathcal{F}_\sigma \subset \mathcal{F}_{\sigma_n}$, then

$$\int_A X_{\tau_n} dP \leq \int_A X_{\sigma_n} dP \qquad (*)$$

As $\sigma_1 \geq \sigma_2 \geq \dots$, by optional sampling th eorem for the countable case, we have

$$E(X_{\sigma_{n-1}} | \mathcal{F}_{\sigma_n}) \leq X_{\sigma_n} .$$

Further

$$\mathcal{F}_{\sigma_1} \supset \mathcal{F}_{\sigma_2} \supset \dots ; E(X_{\sigma_n} | \mathcal{F}_0) \leq X_0 .$$

By the lemma $\{X_{\sigma_n}\}, \{X_{\tau_n}\}$ are uniformly integrable families. By right continuity $X_{\sigma_n} \to X_\sigma$ pointwise and $X_{\tau_n} \to X_\tau$ pointwise.

Letting $n \to \infty$ in (*) we get the required result.

LEMMA. *(Integration by Parts). Let* $M(t,.)$ *be a continuous progressively measurable martingale and* $A(t,w):[0,\infty) \times \Omega \to \mathbb{R}$ *be of bounded variation for each* w. *Further, assume that* $A(t,w)$ *is* $\mathcal{F}_t$ *-measurable for each* t. *Then*

$$Y(t,.) = M(t,.)\, A(t,.) = \int_0^t M(s,.)\, dA(s,.)$$

is a martingale if

$$E\left(\sup_{0 \le s \le t} |M(s,.)|\, \|A(.)\|_t \right) < \infty$$

for each t, *where* $\|A(w)\|_t$ *is the total variation of* $A(s,w)$ *in* $[0,t]$.

Proof. By hypothesis,

$$\sum_{i=0}^{n} M(s\ ,.)\, (A(s_{i+1},.) - A(s_i,.))$$

converges to

$$\int_s^t M(u,.)\, dA(u,.) \quad \text{in } L^1 \text{ as } n \to \infty$$

and as the norm of the partition $s = s_0 < s_1 < .. < s_{n+1} = t$ converges to zero. Hence it is enough to show that

$$E([M(t,.)\, A(t,.) - \sum_{i=0}^{n} M(s_{i+1},.)\, (A(s_{i+1},.) - A(s_i,.))] \mid \mathcal{F}_s)$$

$$= M(s,.)\, A(s,.).$$

But the left side $= E(M(s_{n+1},.)\, A(s_{n+1},.) -$

$$- \sum_{i=0}^{n} M(s_{i+1},.)(A(s_{i+1},.) - A(s_i,.)) \mid \mathcal{F}_s)$$

$$= M(s,.)\, A(s,.).$$

Taking limits as $n \to \infty$ and observing that

$$\sup_{0 \leq i \leq n} |(s_{i+1} - s_i)| \to 0,$$

we get

$$E(M(t,.)\, A(t,.) - \int_0^t M(u,.)\, dA(u,.) | \mathcal{F}_s)$$

$$= M(s,.)\, A(s,.) - \int_0^s M(u,.)\, dA(u,.).$$

----

gk.

BIBLIOGRAPHY

1. BILLINGSLEY, P. *Convergence of probability measures*, John Wiley, New York, (1968).

2. DOOB, J.L. *Stochastic Processes*, John Wiley, New York (1952).

3. DYNKIN, E.B. Markov Processes and Problems in Analysis, *Proc. Int. Cong of Math. Stockholm*, (1962) 36 - 58.

4. DYNKIN, E.B. *Markov Processes*, Vols. 1,2, Springer-Verlag, Berlin (1965).

5. FRIEDMAN, AVNER. *Stochastic Differential Equations and Applications*, Vols. 1,2, Academic Press, New York (1976).

6. FRIEDMAN, AVNER. *Partial Differential Equations of Parabolic type*, Prentice Hall, Englewood Cliffs (1964).

7. GIHMAN, I.I. and A.V. SKOROHOD. *Stochastic Differential Equations*, Springer-Verlag, New York (1973).

8. ITÔ, K. On Stochastic Differential Equations, *Memoirs Amer. Math. Soc. 4* (1951).

9. ITÔ, K. and H.P. McKEAN, Jr. *Diffusion Processes and their sample paths*, Springer-Verlag, Berlin (1965).

10. KUNITA, H. and S. WATANABE. On Square Integrable Martingales, *Nagoya Math. Journal.* 30 (1967) 209 - 245.

11. LÉVY, P. *Processus Stochastic et Mouvement Brownien*, Gaulhier-Villars, Paris (1948).

12. MEYER, P.A. *Probability and Potentials*, Blaisdell, Wallham (1966).

13. NEVEU, J. *Discrete Parameter Martingales*, North Holland, Amsterdam (1975).

14. PARTHASARATHY, K.R. *Probability Measures on Metric Spaces*, Academic Press, New York (1967).

15. PROTTER, M.H. and H.F. WEINBERGER. *Maximum Principles in Differential Equations*, Printice Hall, Englewood Cliffs (1967).

16. STROOCK, D.W. and S.R.S. VARADHAN. *Diffusion Process in Several Dimension*, Springer-Verlag, New York (to appear)

17. WILLIAMS, D. Brownian motions and diffusions as Markov Processes. *Bull. Lond. Math. Soc.* 6 (1974) 257 - 303.

TATA INSTITUTE OF FUNDAMENTAL RESEARCH

LECTURES ON MATHEMATICS AND PHYSICS

General Editors: K.G. RAMANATHAN and B.V. SREEKANTAN

MATHEMATICS

1. On the Riemann Zeta-function. By K. Chandrasekharan*

2. On Analytic Number Theory. By H. Rademacher*

3. On Siegel's Modular Functions. By H. Maass*

4. On Complex Analytic Manifolds. By L. Schwartz*

5. On the Algebraic Theory of Fields. By K.G. Ramanathan*

6. On Sheaf Theory. By C.H. Dowker*

7. On Quadratic Forms. By C.L. Siegel*

8. On Semi-group Theory and its application to Cauchy's problem in Partial Differential Equations. By K. Yosida*

9. On Modular Correspondences. By M. Eichler*

10. On Elliptic Partial Differential Equations. By J.L. Lions*

11. On Mixed Problems in Partial Differential Equations and Representations of Semi-groups. By L. Schwartz*

12. On Measure Theory and Probability. By H.R. Pitt*

13. On the Theory of Functions of Several Complex Variables. By B. Malgrange*

14. On Lie groups and Representations of Locally Compact Groups. By F. Bruhat*

15. On Mean Periodic Functions. By J.P. Kahane*

16. On Approximation by Polynomials. By J.C. Burkill*

17. On Meromorphic Functions. By W.K. Hayman*

---

* Not available

18. On the Theory of Algebraic Functions of One Variable. By M. Deuring*

19. On Potential Theory. By M. Brelot*

20. On Fibre Bundles and Differential Geometry. By J.L. Koszul*

21. On Topics in the Theory of Infinite Groups. By B.H. Neumann*

22. On Topics in Mean Periodic Functions and the Two-Radius Theorem. By J. Delsarte*

23. On Advanced Analytic Number Theory. By C.L. Siegel*

24. On Stochastic Processes. By K. Ito*

25. On Exterior Differential Systems. By M. Kuranishi*

26. On Some Fixed Point Theorems of Functional Analysis. By F.F. Bonsall*

27. On Some Aspects of p-adic Analysis. By F. Bruhat*

28. On Riemann Matrices. By C.L. Siegel*

29. On Modular Functions of one complex variable. By H. Maass

30. On Unique Factorization Doamins. By P. Samuel*

31. On The Fourteenth Problem of Hilbert. By M. Nagata

32. On Groups of Transformations. By J.L. Koszul

33. On Geodesics in Riemannian Geometry. By M. Berger*

34. On Topics in Analysis. By Raghavan Narasimhan*

35. On Cauchy Problem. By Sigeru Mizohata

36. On Old and New Results on Algebraic Curves. By P. Samuel*

37. On Minimal Models and Birational Transformations of Two Dimensional Schemes. By I.R. Shafarevich

38. On Stratification of Complex Analytic sets. By M.-H. Schwartz

39. On Levi Convexity of Complex Manifolds and Cohomology Vanishing Theorems. By E. Vesentini

40. On An Introduction to Grothendieck's Theory of the Fundamental Group. By J.P. Murre

41. On Topics in Algebraic K-Theory. By Hyman Bass*

42. On the Singularities of the three-body problem. By C.L. Siegel

43. On Polyhedral Topology. By J.R. Stallings*

44. On Introduction to Algebraic Topology. By G. de Rham

45. On Quadratic Jordan Algebras. By N. Jacobson*

46. On The Theorem of Browder and Novikov and Siebenmann's Thesis By M.A. Kervaire

47. On Galois Cohomology of Classical Groups. By M. Kneser*

48. On Discrete subgroups of Lie Groups. By G.D. Mostow

49. On the Finite Element Method. By Ph. Ciarlet

50. On Disintegration of measures. By L. Schwartz

51. On Introduction to moduli problems and orbit spaces. By P.E. Newstead

52. On Numerical methods for time dependent equations - Application to Fluid flow problems. By P. Lascaux

53. On Optimisation - Theory and Algorithms. By Jean Cea

54. On Deformations of Singularities. By M. Artin

55. On Sieve Methods. By Hans E. Richert

56. On Irregularities of Distribution. By Wolfgang M. Schmidt

57. On Expansion techniques in Algebraic Geometry. By S.S. Abhyankar

58. On Torus Embeddings and Applications. By T. Oda

59. On Forms of Higher degree. By Jun.-Ichi Igusa

60. On Curves on rational and unirational surfaces. By M. Miyanishi

61. On Wave Propagation. By G.B. Whitham

62. On Equations defining space curves. By L. Szpiro

63. On Topics in Finite element solution of Elliptic Problems. By Bertrand Mercier

64. On Diffusion Problems and partial differential equations. By S.R.S. Varadhan